CLIMATE CHANGE IMPACTS ON THE UNITED STATES

The Potential Consequences of Climate Variability and Change

Overview

Humanity's influence on the global climate will grow in the coming century. Increasingly, there will be significant climate-related changes that will affect each one of us.

We must begin now to consider our responses, as the actions taken today will affect the quality of life for us and future generations.

A Report of the National Assessment Synthesis Team

US Global Change Research Program

This report was produced by the National Assessment Synthesis Team, an advisory committee chartered under the Federal Advisory Committee Act to help the US Global Change Research Program fulfill its mandate under the Global Change Research Act of 1990. The National Science and Technology Council has forwarded this report to the President and Congress for their consideration as required by the Global Change Research Act.

Administrative support for the US Global Change Research Program is provided by the University Corporation for Atmospheric Research, which is sponsored by the National Science Foundation. Any opinions, findings and conclusions or recommendations expressed in this publication are those of the authors and do not necessarily reflect the views of the National Science Foundation or the University Corporation for Atmospheric Research.

The recommended citation of this report is as follows:

National Assessment Synthesis Team,
*Climate Change Impacts on the United States:
The Potential Consequences of Climate Variability and Change,*
US Global Change Research Program, Washington DC, 2000

ABOUT THIS DOCUMENT

What is this Assessment?

The National Assessment of the Potential Consequences of Climate Variability and Change is a landmark in the major ongoing effort to understand what climate change means for the US. Climate science is developing rapidly and scientists are increasingly able to project some changes at the regional scale, identifying regional vulnerabilities, and assessing potential regional impacts. Science increasingly indicates that the Earth's climate has changed in the past and continues to change, and that even greater climate change is very likely in the 21st century. This Assessment has begun a national process of research, analysis, and dialogue about the coming changes in climate, their impacts, and what Americans can do to adapt to an uncertain and continuously changing climate. This Assessment is built on a solid foundation of science conducted as part of the United States Global Change Research Program (USGCRP).

What is this document and who is the NAST?

This document is the Assessment Overview, written by the National Assessment Synthesis Team (NAST). The NAST is a committee of experts drawn from governments, universities, industry, and non-governmental organizations. It has been responsible for broad oversight of the Assessment, with the Federal agencies of the USGCRP. This Overview is based on a longer, referenced "Foundation" report, written by the NAST in cooperation with independent regional and sector assessment teams. These two national-level, peer-reviewed documents synthesize results from studies conducted by regional and sector teams, and from the broader scientific literature.

Why was this Assessment undertaken?

The Assessment was called for by a 1990 law, and has been conducted under the USGCRP in response to a request from the President's Science Advisor. The NAST developed the Assessment's plan, which was then approved by the National Science and Technology Council, the cabinet-level body of agencies responsible for scientific research, including global change research, in the US government.

National Assessment Synthesis Team Members

Jerry M.Melillo,Co-chair
Ecosystems Center
Marine Biological Laboratory

Anthony C. Janetos,Co-chair
World Resources Institute

Thomas R.Karl,Co-chair
NOAA National Climatic Data Center

Robert Corell (from January 2000)
American Meteorological Society and
Harvard University

Eric J. Barron
Pennsylvania State University

Virginia Burkett
USGS National Wetlands Research Center

Thomas F. Cecich
Glaxo Wellcome Inc.

Katharine Jacobs
Arizona Department of Water Resources

Linda Joyce
USDA Forest Service

Barbara Miller
World Bank

M.Granger Morgan
Carnegie Mellon University

Edward A. Parson (until January 2000)
Harvard University

Richard G. Richels
EPRI

David S.Schimel
National Center for Atmospheric Research

Additional Lead Authors
Lynne Carter (National
  Assessment Coordination Office)
David Easterling (NOAA National
  Climatic Data Center)
Benjamin Felzer (National Center for
  Atmospheric Research)
John Field (University of Washington)
Paul Grabhorn (Grabhorn Studio)
Susan Joy Hassol (Aspen Global
  Change Institute)
Michael MacCracken (National
  Assessment Coordination Office)
Joel Smith (Stratus Consulting)
Melissa Taylor (National Assessment
  Coordination Office)
Thomas Wilbanks (Oak Ridge National
  Laboratory)

Independent Review Board of the President's Committee of Advisers on Science and Technology (PCAST)

Peter Raven,Co-chair
Missouri Botanical Garden and PCAST

Mario Molina,Co-chair
MIT and PCAST

Burton Richter
Stanford University

Linda Fisher
Monsanto

Kathryn Fuller
World Wildlife Fund

John Gibbons
National Academy of Engineering

Marcia McNutt
Monterey Bay Aquarium Research Institute

Sally Ride
University of California San Diego and PCAST

William Schlesinger
Duke University

James Gustave Speth
Yale University

Robert White
University Corporation for Atmospheric
Research,and Washington Advisory Group

Other Contributors

This report is also based on the work of hundreds of individuals and organizations participating in regional and sector activities across the country without whose input,support, and expertise,it would not have been possible.In addition,many reviewers provided comments on drafts of the report.Additional credits and acknowledgements can be found in the appendix.

TABLE OF CONTENTS

ABOUT THE ASSESSMENT PROCESS

What is the purpose of this Assessment?

The Assessment's purpose is to synthesize, evaluate, and report on what we presently know about the potential consequences of climate variability and change for the US in the 21st century. It has sought to identify key climatic vulnerabilities of particular regions and sectors, in the context of other changes in the nation's environment, resources, and economy. It has also sought to identify potential measures to adapt to climate variability and change. Finally, because present knowledge is limited, the Assessment has sought to identify the highest priority uncertainties about which we must know more to understand climate impacts, vulnerabilities, and our ability to adapt.

How did the process involve both stakeholders and scientists in this Assessment?

This first National Assessment involved both stakeholders and scientific experts. Stakeholders included, for example, public and private decision-makers, resource and environmental managers, and the general public. The stakeholders from different regions and sectors began the Assessment by articulating their concerns in a series of workshops about climate change impacts in the context of the other major issues they face. In the workshops and subsequent consultations, stakeholders identified priority regional and sector concerns, mobilized specialized expertise, identified potential adaptation options, and provided useful information for decision-makers. The Assessment also involved many scientific experts using advanced methods, models, and results. Further, it has stimulated new scientific research in many areas and identified priority needs for further research.

What is the breadth of this Assessment?

Although global change embraces many interrelated issues, this first National Assessment has examined only climate change and variability, with a primary focus on specific regions and sectors. In some cases, regional and sector analyses intersect and complement each other. For example, the Forest sector and the Pacific Northwest have both provided insights into climate impacts on Northwest forests.

The regions cover the nation. Impacts outside the US are considered only briefly, with particular emphasis on potential linkages to the US. Sector teams examined Water, Agriculture, Human Health, Forests, and Coastal Areas and Marine Resources. This first Assessment could not attempt to be comprehensive: the choice of these five sectors reflected an expectation that they were likely to be both important and particularly informative, and that relevant data and analytic tools were available – not a conclusion that they are the only important domains of climate impact. Among the sectors considered, there was a continuum in the amount of information available to support the Assessment, with some being at far earlier stages of development. Future assessments should consider other potentially important issues, such as Energy, Transportation, Urban Areas, and Wildlife.

Each regional and sector team is publishing a separate report of its own analyses, some of which are still continuing. The Overview and Foundation reports consequently represent a snapshot of our understanding at the present time.

After identifying potential impacts of climate change, what kinds of societal responses does this report explore?

Responses to climate change can be of two broad types. One type involves adaptation measures to reduce the harms and risks, and maximize the benefits and opportunities, of climate change, whatever its cause. The other type involves mitigation measures to reduce human contributions to climate change. After identifying potential impacts, this Assessment sought to identify potential adaptation measures for each region and sector studied. While this was an important first step, it was not possible at this stage to evaluate the practicality, effectiveness, or costs of the potential adaptation measures. Both mitigation and adaptation measures are necessary elements of a coherent and integrated response to climate change. Mitigation measures were not included in this Assessment, but are being assessed in other bodies such as the United Nations Intergovernmental Panel on Climate Change (IPCC).

Does the fact that this report excludes mitigation mean that nothing can be done to reduce climate change?

No. An integrated climate policy will combine mitigation and adaptation measures as appropriate. If future world emissions of greenhouse gases are lower than currently projected, for whatever reason, including intentional mitigation, then the rate of climate change, the associated impacts, and the cost and difficulty of adapting will all be reduced. If emissions are higher than expected, then the rate of change, the impacts, and the difficulty of adapting will be increased. But no matter how aggressively emissions are reduced, the world will still experience at least a century of climate change. This will happen because the elevated concentrations of greenhouse gases already in the atmosphere will remain for many decades, and because the climate system responds to changes in human inputs only very slowly. Consequently, even if the world takes mitigation measures, we must still adapt to a changing climate. Similarly, even if we take adaptation measures, future emissions will have to be curbed to stabilize climate. Neither type of response can completely supplant the other.

How are computer models used in this Assessment?

State-of-the-science climate models have been used to generate climate change scenarios. Computer models of ecological systems, hydrological systems, and various socioeconomic systems have also been used in the Assessment, to study responses of these systems to the scenarios generated by climate models.

What additional tools, besides models, were used to evaluate potential climate change impacts?

In addition to models, the Assessment has used two other ways to think about potential future climate. First, the Assessment has used historical climate records to evaluate sensitivities of regions and sectors to climate variability and extremes that have occurred in the 20th century. Looking at real historical climate events, their impacts, and how people have adapted, gives valuable insights into potential future impacts that complement those provided by model projections. In addition, the Assessment has used sensitivity analyses, which ask how, and how much, the climate would have to change to bring major impacts on particular regions or sectors. For example, how much would temperature have to increase in the South before agricultural crops such as soybeans would be negatively affected? What would be the result for forest productivity of continued increases in temperature and leveling off of the CO_2 fertilization effect?

Has this report been peer reviewed?

This Overview and the underlying Foundation document have been extensively reviewed. More than 300 scientific and technical experts have provided detailed comments on part or all of the report in two separate technical reviews. The report was reviewed at each stage for technical accuracy by the agencies of the US Global Change Research Program. The public also provided hundreds of helpful suggestions for clarification and modification during a 60-day public comment period. A panel of distinguished experts convened by the President's Committee of Advisors on Science and Technology has provided broad oversight, and monitored the authors response to all reviews.

ABOUT SCENARIOS AND UNCERTAINTY

What are scenarios and why are they used?

Scenarios are plausible alternative futures – each an example of what might happen under particular assumptions. Scenarios are not specific predictions or forecasts. Rather, scenarios provide a starting point for examining questions about an uncertain future and can help us visualize alternative futures in concrete and human terms. The military and industry frequently use these powerful tools for future planning in high-stakes situations. Using scenarios helps to identify vulnerabilities and plan for contingencies.

Why are climate scenarios used in this Assessment and how were they developed?

Because we cannot predict many aspects of our nation's future climate, we have used scenarios to help explore US vulnerability to climate change. Results from state-of-the-science climate models and data from historical observations have been used to generate a variety of such scenarios. Projections of changes in climate from the Hadley Centre in the United Kingdom and the Canadian Centre for Climate Modeling and Analysis served as the primary resources for this Assessment. Results were also drawn from models developed at the National Center for Atmospheric Research, NOAA's Geophysical Fluid Dynamics Laboratory, and NASA's Goddard Institute for Space Studies.

For some aspects of climate, virtually all models, as well as other lines of evidence, agree on the types of changes to be expected. For example, all climate models suggest that the climate is going to get warmer, the heat index is going to rise, and precipitation is more likely to come in heavy and extreme events. This consistency lends confidence to these results.

For some other aspects of climate, however, the model results differ. For example, some models, including the Canadian model, project more extensive and frequent drought in the US, while others, including the Hadley model, do not. The Canadian model suggests a drier Southeast in the 21st century while the Hadley model suggests a wetter one. In such cases, the scenarios provide two plausible but different alternatives. Such differences can help identify areas in which the models need improvement.

Many of the maps in this document are derived from the two primary climate model scenarios. In most cases, there are three maps: one shows average conditions based on actual observations from 1961-1990; the other two are generated by the Hadley and Canadian model scenarios and reflect the model's projection of change from those average conditions.

What assumptions about emissions are in these two climate scenarios?

Because future trends in fossil fuel use and other human activities are uncertain, the Intergovernmental Panel on Climate Change (IPCC) has developed a set of scenarios for how the 21st century may evolve. These scenarios consider a wide range of possibilities for changes in population, economic growth, technological development, improvements in energy efficiency, and the like. The two primary climate scenarios used in this Assessment are based on one mid-range emissions scenario for the future that assumes no major changes in policies to limit greenhouse gas emissions. Some other important assumptions in this scenario are that by the year 2100:

- world population will nearly double to about 11 billion people;
- the global economy will continue to grow at about the average rate it has been growing, reaching more than ten times its present size;
- increased use of fossil fuels will triple CO_2 emissions and raise sulfur dioxide emissions, resulting in an atmospheric CO_2 concentration of just over 700 parts per million; and
- total energy produced each year from non-fossil sources such as wind, solar, biomass, hydroelectric, and nuclear will increase to more than ten times its current amount, providing more than 40% of the world's energy, rather than the current 10%.

Many of the maps in this document are derived from the two primary climate model scenarios. In most cases, there are three maps: one shows average conditions based on actual observations from 1961-1990; the other two are generated by the Hadley and Canadian model scenarios and reflect the model's projection of change from those average conditions.

The Assessment's Emissions Scenario Falls in the Middle of the other IPCC Emissions Scenarios

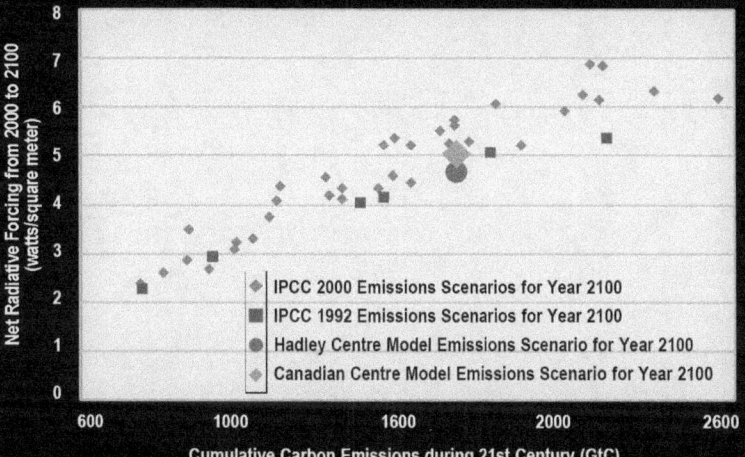

Net Radiative Forcing from 2000 to 2100 (watts/square meter)

Cumulative Carbon Emissions during 21st Century (GtC)

◆ IPCC 2000 Emissions Scenarios for Year 2100
■ IPCC 1992 Emissions Scenarios for Year 2100
● Hadley Centre Model Emissions Scenario for Year 2100
◆ Canadian Centre Model Emissions Scenario for Year 2100

The graph shows a comparison of the projections of total carbon dioxide emissions (in billions of metric tons of carbon, GtC) and the human-induced warming influence due to all the greenhouse gases and sulfate aerosols for the emissions scenarios prepared by the IPCC in 1992 and 2000. As is apparent from the graph, both the emissions scenario and the human-induced warming influence assumed in this Assessment lie near the mid-range of the set of IPCC scenarios. Further detail can be found in the Climate chapter in the Foundation report.

Both the emissions scenario and the human-induced warming influence assumed in this Assessment lie near the mid-range of the set of IPCC scenarios.

How is the likelihood of various impacts expressed?

To integrate a wide variety of information and differentiate more likely from less likely outcomes, the NAST developed a common language to express the team's considered judgement about the likelihood of results. The NAST developed its collective judgements through discussion and consideration of the supporting information. Historical data, model projections, published scientific literature, and other available information all provided input to these deliberations, except where specifically stated that the result comes from a particular model scenario. In developing these judgements, there were often several lines of supporting evidence (e.g., drawn from observed trends, analytic studies, model simulations). Many of these judgements were based on broad scientific consensus as stated by well-recognized authorities including the IPCC and the National Research Council. In many cases, groups outside the NAST reviewed the use of terms to provide input from a broader set of experts in a particular field.

Language Used to Express Considered Judgement

Common Language

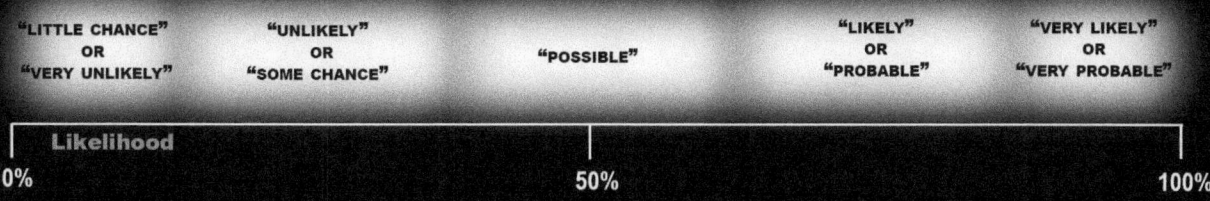

| "LITTLE CHANCE" OR "VERY UNLIKELY" | "UNLIKELY" OR "SOME CHANCE" | "POSSIBLE" | "LIKELY" OR "PROBABLE" | "VERY LIKELY" OR "VERY PROBABLE" |

Likelihood

0%   50%   100%

SUMMARY

CLIMATE CHANGE AND OUR NATION

The findings in this report are based on a synthesis of historical data, model projections, published scientific research, and other available information, except where specifically noted.

Long-term observations confirm that our climate is now changing at a rapid rate. Over the 20th century, the average annual US temperature has risen by almost 1°F (0.6°C) and precipitation has increased nationally by 5 to 10%, mostly due to increases in heavy downpours. These trends are most apparent over the past few decades. The science indicates that the warming in the 21st century will be significantly larger than in the 20th century. Scenarios examined in this Assessment, which assume no major interventions to reduce continued growth of world greenhouse gas emissions, indicate that temperatures in the US will rise by about 5-9°F (3-5°C) on average in the next 100 years, which is more than the projected *global* increase. This rise is very likely to be associated with more extreme precipitation and faster evaporation of water, leading to greater frequency of both very wet and very dry conditions.

This Assessment reveals a number of national-level impacts of climate variability and change including impacts to natural ecosystems and water resources. Natural ecosystems appear to be the most vulnerable to the harmful effects of climate change, as there is often little that can be done to help them adapt to the projected speed and amount of change. Some ecosystems that are already constrained by climate, such as alpine meadows in the Rocky Mountains, are likely to face extreme stress, and disappear entirely in some places. It is likely that other more widespread ecosystems will also be vulnerable to climate change. One of the climate scenarios used in this Assessment suggests the potential for the forests of the Southeast to break up into a mosaic of forests, savannas, and grasslands. Climate scenarios suggest likely changes in the species composition of the Northeast forests, including the loss of sugar maples. Major alterations to natural ecosystems due to climate change could possibly have negative consequences for our economy, which depends in part on the sustained bounty of our nation's lands, waters, and native plant and animal communities.

A unique contribution of this first US Assessment is that it combines national-scale analysis with an examination of the potential impacts of climate change on different regions of the US. For example, sea-level rise will very likely cause further loss of coastal wetlands (ecosystems that provide vital nurseries and habitats for many fish species) and put coastal communities at greater risk of storm surges, especially in the Southeast. Reduction in snowpack will very likely alter the timing and amount of water supplies, potentially exacerbating water shortages and conflicts, particularly throughout the western US. The melting of glaciers in the high-elevation West and in Alaska represents the loss or diminishment of unique national treasures of the American landscape. Large increases in the heat index (which combines temperature and humidity) and increases in the frequency of heat waves are very likely. These changes will, at minimum, increase discomfort, particularly in cities. It is very probable that continued thawing of permafrost and melting of sea ice in Alaska will further damage forests, buildings, roads, and coastlines, and harm subsistence livelihoods. In various parts of the nation, cold-weather recreation such as skiing will very likely be reduced, and air conditioning usage will very likely increase.

Highly managed ecosystems appear more robust, and some potential benefits have been identified. Crop and forest productivity is likely to increase in some areas for the next few decades due to increased carbon dioxide in the atmosphere and an extended growing season. It is possible that some US food exports could increase, depending on impacts in other food-growing regions around the world. It is also possible that a rise in crop production in fertile areas could cause prices to fall, benefiting consumers. Other benefits that are possible include extended seasons for construction and warm weather recreation, reduced heating requirements, and reduced cold-weather mortality.

Climate variability and change will interact with other environmental stresses and socioeconomic changes. Air and water pollution, habitat fragmentation, wetland loss, coastal erosion, and reductions in fisheries are likely to be compounded by climate-related stresses. An aging populace nationally, and rapidly growing populations in cities, coastal areas, and across the South and West are social factors that interact with and alter sensitivity to climate variability and change.

There are also very likely to be unanticipated impacts of climate change during the next century. Such "surprises" may stem from unforeseen changes in the physical climate system, such as major alterations in ocean circulation, cloud distribution, or storms; and unpredicted biological consequences of these physical climate changes, such as massive dislocations of species or pest outbreaks. In addition, unexpected social or economic change, including major shifts in wealth, technology, or political priorities, could affect our ability to respond to climate change.

Greenhouse gas emissions lower than those assumed in this Assessment would result in reduced impacts. The signatory nations of the Framework Convention on Climate Change are negotiating the path they will ultimately take. Even with such reductions, however, the planet and the nation are certain to experience more than a century of climate change, due to the long lifetimes of greenhouse gases already in the atmosphere and the momentum of the climate system. Adapting to a changed climate is consequently a necessary component of our response strategy.

The warming in the 21st century will be significantly larger than in the 20th century.

Natural ecosystems, which are our life support system in many important ways, appear to be the most vulnerable to the harmful effects of climate change...

Major alterations to natural ecosystems due to climate change could possibly have negative consequences for our economy, which depends in part on the sustained bounty of our nation's lands, waters, and native plant and animal communities.

SUMMARY

CLIMATE CHANGE AND OUR NATION

The magnitude of climate change impacts depends on time period and geographic scale. Short-term impacts differ from long-term impacts, and regional and local level impacts are much more pronounced than those at the national level.

For the nation as a whole, direct economic impacts are likely to be modest, while in some places, economic losses or gains are likely to be large. For example, while crop yields are likely to increase at the national scale over the next few decades, large increases or decreases in yields of specific crops in particular places are likely.

Through time, climate change will possibly affect the same resource in opposite ways. For example, forest productivity is likely to increase in the short term, while over the longer term, changes in processes such as fire, insects, drought, and disease will possibly decrease forest productivity.

A daptation measures can, in many cases, reduce the magnitude of harmful impacts, or take advantage of beneficial impacts. For example, in agriculture, many farmers will probably be able to alter cropping and management practices. Roads, bridges, buildings, and other long-lived infrastructure can be designed taking projected climate change into account. Adaptations, however, can involve trade-offs, and do involve costs. For example, the benefits of building sea walls to prevent sea-level rise from disrupting human coastal communities will need to be weighed against the economic and ecological costs of seawall construction. The ecological costs could be high as seawalls prevent the inland shifting of coastal wetlands in response to sea-level rise, resulting in the loss of vital fish and bird habitat and other wetland functions, such as protecting shorelines from damage due to storm surges. Protecting against any increased risk of water-borne and insect-borne diseases will require diligent maintenance of our public health system. Many adaptations, notably those that seek to reduce other environmental stresses such as pollution and habitat fragmentation, will have beneficial effects beyond those related to climate change.

Vulnerability in the US is linked to the fates of other nations, and we cannot evaluate national consequences due to climate variability and change without also considering the consequences of changes elswhere in the world. The US is linked to other nations in many ways, and both our vulnerabilities and our potential responses will likely depend in part on impacts and responses in other nations. For example, conflicts or mass migrations resulting from resource limits, health, and environmental stresses in more vulnerable nations could possibly pose challenges for global security and US policy. Effects of climate variability and change on US agriculture will depend critically on changes in agricultural productivity elsewhere, which can shift international patterns of food supply and demand. Climate-induced changes in water resources available for power generation, transportation, cities, and agriculture are likely to raise potentially delicate diplomatic issues with both Canada and Mexico.

This Assessment has identified many remaining uncertainties that limit our ability to fully understand the spectrum of potential consequences of climate change for our nation. To address these uncertainties, additional research is needed to improve understanding of ecological and social processes that are sensitive to climate, application of climate scenarios and reconstructions of past climates to impacts studies, and assessment strategies and methods. Results from these research efforts will inform future assessments that will continue the process of building our understanding of humanity's impacts on climate, and climate's impacts on us.

KEY FINDINGS

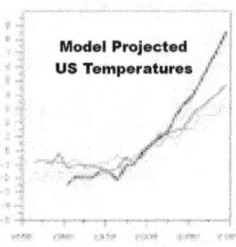
Model Projected US Temperatures

1. Increased warming

Assuming continued growth in world greenhouse gas emissions, the primary climate models used in this Assessment project that temperatures in the US will rise 5-9°F (3-5°C) on average in the next 100 years. A wider range of outcomes is possible.

2. Differing regional impacts

Climate change will vary widely across the US. Temperature increases will vary somewhat from one region to the next. Heavy and extreme precipitation events are likely to become more frequent, yet some regions will get drier. The potential impacts of climate change will also vary widely across the nation.

3. Vulnerable ecosystems

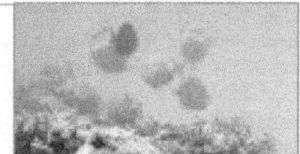

Many ecosystems are highly vulnerable to the projected rate and magnitude of climate change. A few, such as alpine meadows in the Rocky Mountains and some barrier islands, are likely to disappear entirely in some areas. Others, such as forests of the Southeast, are likely to experience major species shifts or break up into a mosaic of grasslands, woodlands, and forests. The goods and services lost through the disappearance or fragmentation of certain ecosystems are likely to be costly or impossible to replace.

4. Widespread water concerns

Water is an issue in every region, but the nature of the vulnerabilities varies. Drought is an important concern in every region. Floods and water quality are concerns in many regions. Snowpack changes are especially important in the West, Pacific Northwest, and Alaska.

5. Secure food supply

At the national level, the agriculture sector is likely to be able to adapt to climate change. Overall, US crop productivity is very likely to increase over the next few decades, but the gains will not be uniform across the nation. Falling prices and competitive pressures are very likely to stress some farmers, while benefiting consumers.

6. Near-term increase in forest growth

Forest productivity is likely to increase over the next several decades in some areas as trees respond to higher carbon dioxide levels. Over the longer term, changes in larger-scale processes such as fire, insects, droughts, and disease will possibly decrease forest productivity. In addition, climate change is likely to cause long-term shifts in forest species, such as sugar maples moving north out of the US.

7. Increased damage in coastal and permafrost areas

Climate change and the resulting rise in sea level are likely to exacerbate threats to buildings, roads, powerlines, and other infrastructure in climatically sensitive places. For example, infrastructure damage is related to permafrost melting in Alaska, and to sea-level rise and storm surge in low-lying coastal areas.

8. Adaptation determines health outcomes

A range of negative health impacts is possible from climate change, but adaptation is likely to help protect much of the US population. Maintaining our nation's public health and community infrastructure, from water treatment systems to emergency shelters, will be important for minimizing the impacts of water-borne diseases, heat stress, air pollution, extreme weather events, and diseases transmitted by insects, ticks, and rodents.

9. Other stresses magnified by climate change

Climate change will very likely magnify the cumulative impacts of other stresses, such as air and water pollution and habitat destruction due to human development patterns. For some systems, such as coral reefs, the combined effects of climate change and other stresses are very likely to exceed a critical threshold, bringing large, possibly irreversible impacts.

10. Uncertainties remain and surprises are expected

Significant uncertainties remain in the science underlying regional climate changes and their impacts. Further research would improve understanding and our ability to project societal and ecosystem impacts, and provide the public with additional useful information about options for adaptation. However, it is likely that some aspects and impacts of climate change will be totally unanticipated as complex systems respond to ongoing climate change in unforeseeable ways.

IMPACTS OF CLIMATE CHANGE

It is very likely that the US will get substantially warmer. Temperatures are projected to rise more rapidly in the next one hundred years than in the last 10,000 years. It is also very likely that there will be more precipitation overall, with more of it coming in heavy downpours. In spite of this, some areas are likely to get drier as increased evaporation due to higher temperatures outpaces increased precipitation. Droughts and flash floods are likely to become more frequent and intense.

SPECIES DIVERSITY

While it is possible that some species will adapt to changes in climate by shifting their ranges, human and geographic barriers, and the presence of invasive non-native species will limit the degree of adaptation that can occur. Losses in local biodiversity are likely to accelerate towards the end of the 21st century.

PERMAFROST AREAS

It is very probable that rising temperatures will cause further permafrost thawing, damaging roads, buildings, and forests in Alaska.

FORESTRY

Timber inventories are likely to increase over the 21st century. Hardwood productivity is likely to increase more than softwood productivity in some regions, including the Southeast.

WATER SUPPLY

Reduced summer runoff, increased winter runoff, and increased demands are likely to compound current stresses on water supplies and flood management,especially in the western US.

ISLANDS

Sea-level rise and storm surges will very likely threaten public health and safety and possibly reduce the availability of fresh water.

CORAL REEFS

Increased CO_2 and ocean temperatures,especially combined with other stresses,will possibly exacerbate coral reef bleaching and die-off.

FRESHWATER ECOSYSTEMS

Increases in water temperature and changes in seasonal patterns of runoff will very likely disturb fish habitat and affect recreational uses of lakes,streams, and wetlands.

OUR CHANGING CLIMATE

Climate and the Greenhouse Effect

Global projections based on population growth and assumptions about energy use indicate that the CO_2 concentration will continue to rise, likely reaching somewhere between two and three times its pre-industrial level by 2100.

E arth's climate is determined by complex interactions between the sun, oceans, atmosphere, land, and living things. The composition of the atmosphere is particularly important because certain gases (including water vapor, carbon dioxide, methane, halocarbons, ozone, and nitrous oxide) absorb heat radiated from the Earth's surface. As the atmosphere warms, it in turn radiates heat back to the surface, to create what is commonly called the "greenhouse effect." Changes in the composition of the atmosphere alter the intensity of the greenhouse effect. Such changes, which have occurred many times in the planet's history, have helped determine past climates and will affect the future climate as well.

Human Activities Alter the Balance

H umans are exerting a major and growing influence on some of the key factors that govern climate by changing the composition of the atmosphere and by modifying the land surface. The human impact on these factors is clear. The concentration of carbon dioxide (CO_2) has risen about 30% since the late 1800s. The concentration of CO_2 is now higher than it has been in at least the last 400,000 years. This increase has resulted from the burning of coal, oil, and natural gas, and the destruction of forests around the world to provide space for agriculture and other human activities. Rising concentrations of CO_2 and other greenhouse gases are intensifying Earth's natural greenhouse effect. Global projections of population growth and assumptions about energy use indicate that the CO_2 concentration will continue to rise, likely reaching between two and three times its late-19th-century level by 2100. This dramatic doubling or tripling will occur in the space of about 200 years, a brief moment in geological history.

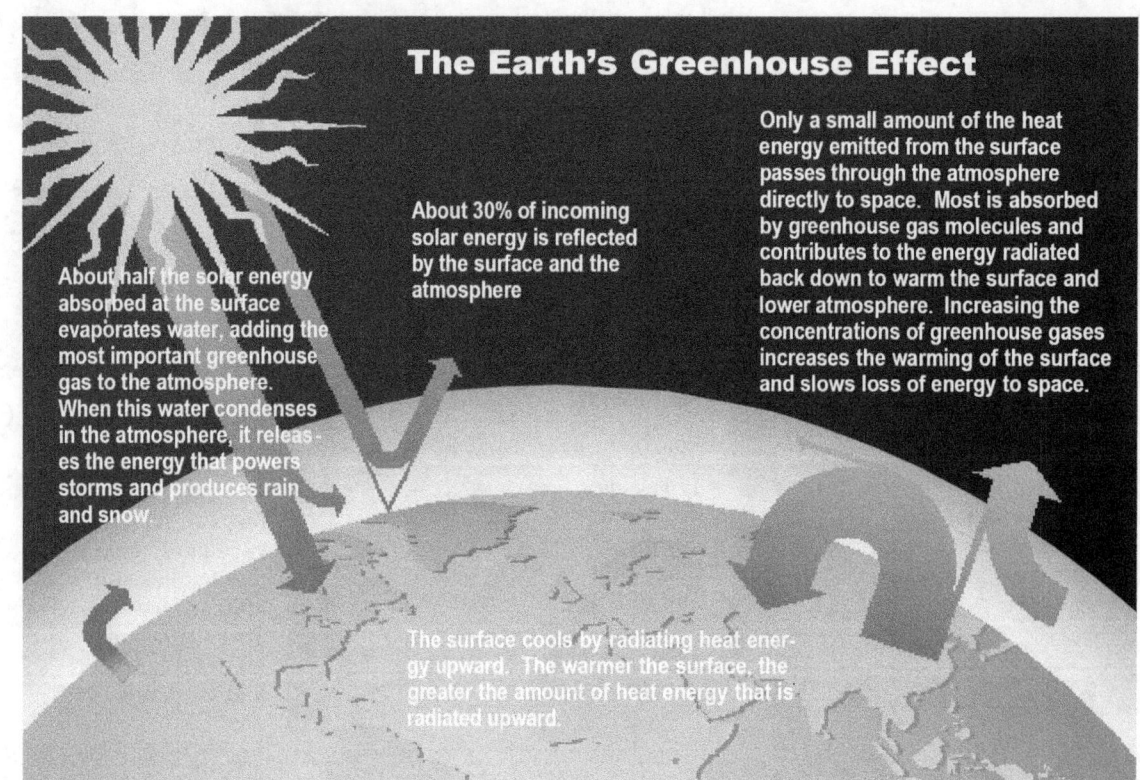

The Earth's Greenhouse Effect

About 30% of incoming solar energy is reflected by the surface and the atmosphere

About half the solar energy absorbed at the surface evaporates water, adding the most important greenhouse gas to the atmosphere. When this water condenses in the atmosphere, it releases the energy that powers storms and produces rain and snow.

Only a small amount of the heat energy emitted from the surface passes through the atmosphere directly to space. Most is absorbed by greenhouse gas molecules and contributes to the energy radiated back down to warm the surface and lower atmosphere. Increasing the concentrations of greenhouse gases increases the warming of the surface and slows loss of energy to space.

The surface cools by radiating heat energy upward. The warmer the surface, the greater the amount of heat energy that is radiated upward.

The Climate Is Changing

As we add more CO_2 and other heat-trapping gases to the atmosphere, the world is becoming warmer (which changes other aspects of climate as well). Historical records of temperature and precipitation have been extensively analyzed in many scientific studies. These studies demonstrate that the global average surface temperature has increased by over 1°F (0.6°C) during the 20th century. About half this rise has occurred since the late 1970s. Seventeen of the eighteen warmest years in the 20th century occurred since 1980. In 1998, the global temperature set a new record by a wide margin, exceeding that of the previous record year, 1997, by about 0.3°F (0.2°C). Higher latitudes have warmed more than equatorial regions, and nighttime temperatures have risen more than daytime temperatures.

As the Earth warms, more water evaporates from the oceans and lakes, eventually to fall as rain or snow. During the 20th century, annual precipitation has increased about 10% in the mid- and high-latitudes. The warming is also causing permafrost to thaw, and is melting sea ice, snow cover, and mountain glaciers. Global sea level rose 4 to 8 inches (10-20 cm) during the 20th century because ocean water expands as it warms and because melting glaciers are adding water to the oceans.

Records of Northern Hemisphere surface temperatures, CO_2 concentrations, and carbon emissions show a close correlation. **Temperature Change:** reconstruction of annual-average Northern Hemisphere surface air temperatures derived from historical records, tree rings, and corals (blue), and air temperatures directly measured (purple). **CO_2 Concentrations:** record of global CO_2 concentration for the last 1000 years, derived from measurements of CO_2 concentration in air bubbles in the layered ice cores drilled in Antarctica (blue line) and from atmospheric measurements since 1957. **Carbon Emissions:** reconstruction of past emissions of CO_2 as a result of land clearing and fossil fuel combustion since about 1750 (in billions of metric tons of carbon per year).

According to the Intergovernmental Panel on Climate Change (IPCC), scientific evidence confirms that human activities are a discernible cause of a substantial part of the warming experienced over the 20th century. New studies indicate that temperatures in recent decades are higher than at any time in at least the past 1,000 years. It is very unlikely that these unusually high temperatures can be explained solely by natural climate variations.

The intensity and pattern of temperature changes within the atmosphere implicates human activities as a cause.

The relevant question is not whether the increase in greenhouse gases is contributing to warming, but rather, what will be the amount and rate of future warming and associated climate changes, and what impacts will those changes have on human and natural systems.

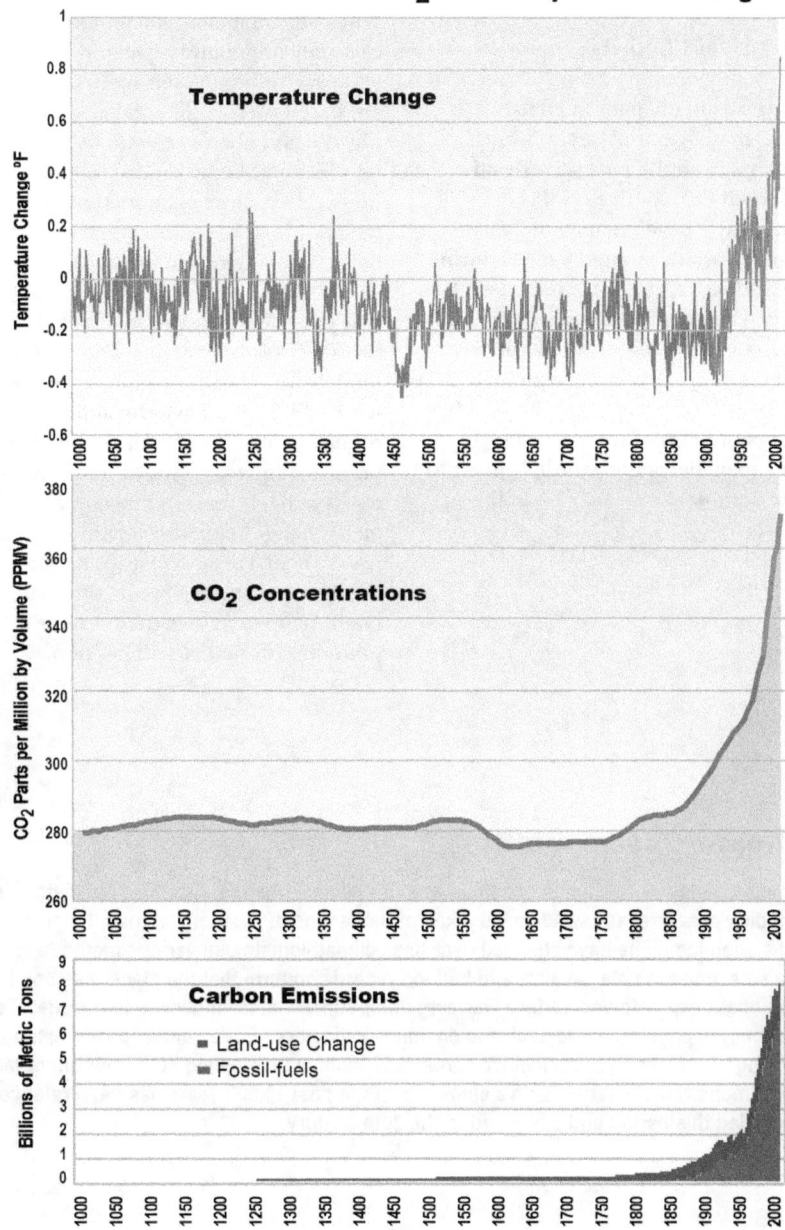

1000 Years of Global CO_2 and Temperature Change

TOOLS FOR ASSESSING CLIMATE CHANGE IMPACTS

For this study, three tools were used to examine the potential impacts of climate change on the US: historical records, comprehensive state-of-the-science climate simulation models, and sensitivity analyses designed to explore our vulnerability to future climate change. These three tools were used because prudent risk management requires consideration of a spectrum of possibilities.

Historical Records

How do changes in climate affect human and natural systems? Records from the past provide an informed perspective on this question. There have been a number of climate variations and changes during the 20th century. These include substantial warming, increases in precipitation, decade-long droughts, and reduction in snow cover extent. Analyzing these variations, and their effects on human and natural systems, provides important insights into how vulnerable we may be in the future.

Climate Model Simulations

Although Earth's climate is astoundingly complex, our ability to use supercomputers to simulate the climate is growing. Today's climate models are not infallible, but they are powerful tools for understanding what the climate might be like in the future.

A key advantage of climate models is that they are quantitative and grounded in scientific measurements. They are based on fundamental laws of physics and chemistry, and incorporate human and biological interactions. They allow examination of a range of possible futures that cannot be examined experimentally.

Our confidence in the accuracy of climate models is growing. The best models have been carefully evaluated by the IPCC and have the ability to replicate most aspects of past and present climates. Two of these models have been used to develop climate change scenarios for this Assessment. These scenarios should be regarded as projections of what might happen, rather than precise predictions of what will happen.

Sensitivity Analyses

What degree of climate change would cause significant impacts to natural and human systems? In other words, how vulnerable and adaptable are we? To help answer such questions, scientists can perform "sensitivity analyses" to determine under what conditions and to what degree a system is sensitive to change. Such analyses are not predictions that such changes will, in fact, occur; rather, they examine what the implications would be if the specified changes did occur. For example, an analyst might ask, "How large would climate change have to be in order to cause a specified impact?"

Climate Observations

Climatologists use two types of data to monitor climate change. The first are historical measurements of temperature, precipitation, humidity, pressure, and wind speed taken at thousands of locations across the globe. Because observing methods, instruments, and station locations have changed over time, climatologists use various methods to crosscheck and corroborate these historical data sets. For example, satellite and balloon records confirm that the planet has been warming for the past four decades, although rates of atmospheric and surface warming differ somewhat from decade to decade. To peer further back into the past, climatologists also analyze physical, biological, and chemical indicators. For example, past climate conditions can be inferred from the width of tree rings, air trapped in ancient ice cores, and sediment deposited at the bottom of lakes and oceans. Taken together, this information demonstrates that the Earth's climate over the past 10,000 years has been relatively stable compared to the 10,000 years that preceeded this period and compared to the 20th century.

Modeling the Climate System

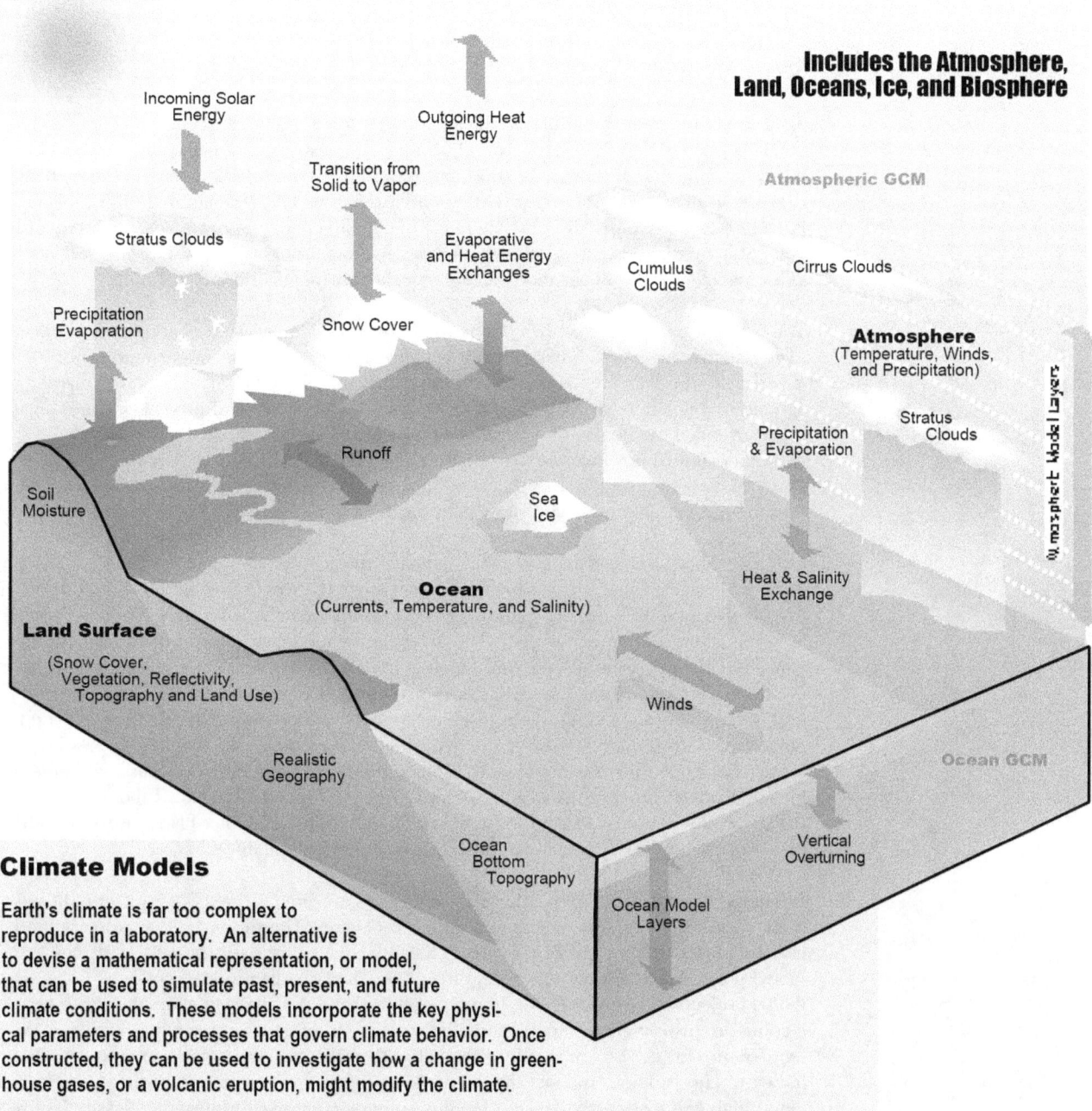

Includes the Atmosphere, Land, Oceans, Ice, and Biosphere

Incoming Solar Energy

Outgoing Heat Energy

Transition from Solid to Vapor

Atmospheric GCM

Stratus Clouds

Evaporative and Heat Energy Exchanges

Cumulus Clouds

Cirrus Clouds

Precipitation Evaporation

Snow Cover

Atmosphere
(Temperature, Winds, and Precipitation)

Precipitation & Evaporation

Stratus Clouds

Runoff

Soil Moisture

Sea Ice

Atmosphere Model Layers

Heat & Salinity Exchange

Ocean
(Currents, Temperature, and Salinity)

Land Surface

(Snow Cover, Vegetation, Reflectivity, Topography and Land Use)

Winds

Realistic Geography

Ocean GCM

Ocean Bottom Topography

Vertical Overturning

Ocean Model Layers

Climate Models

Earth's climate is far too complex to reproduce in a laboratory. An alternative is to devise a mathematical representation, or model, that can be used to simulate past, present, and future climate conditions. These models incorporate the key physical parameters and processes that govern climate behavior. Once constructed, they can be used to investigate how a change in greenhouse gases, or a volcanic eruption, might modify the climate.

Computer models that simulate Earth's climate are called General Circulation Models or GCMs. The models can be used to simulate changes in temperature, rainfall, snow cover, winds, soil moisture, sea ice, and ocean circulation over the entire globe through the seasons and over periods of decades. However, mathematical models are obviously simplified versions of the real Earth that cannot capture its full complexity, especially at smaller geographic scales. Real uncertainties remain in the ability of models to simulate many aspects of the future climate. The models provide a view of future climate that is physically consistent and plausible, but incomplete. Nonetheless, through continual improvement over the last several decades, today's GCMs provide a state-of-the-science glimpse into the next century to help understand how climate change may affect the nation.

TOOLS FOR ASSESSING CLIMATE CHANGE IMPACTS

Scenarios of the Future

Information about the future is valuable, even if it is somewhat uncertain. For example, many people plan their days around weather forecasts with uncertainty conveyed in words or numbers. If there is "a 70% chance of rain" we might take an umbrella with us to work. It may not rain, but if it does, we are prepared. Likewise, although the tools used in this report to explore the possible range of climate change impacts – historical records, computer simulations, and sensitivity analyses – contain uncertainties, their use still provides much valuable information for policymakers, planners, and citizens.

The fact that the climate is changing is apparent from detailed historical records of climate that provide a benchmark for assessing the future. Scientists' understanding of America's future climate – and of the impacts that this altered climate is likely to have on agriculture, human health, water resources, natural ecosystems, and other key issues – has been advanced by the use of computer simulations. Together, the historical record and computer simulations indicate that America's climate is very likely to continue changing in the 21st century, and indeed, that these changes are likely to be substantially larger than those in the 20th century, with significant impacts on our nation.

Climate Models used in the US Assessment

Climate models continue to improve, and assumptions about future greenhouse gas emissions continue to evolve. The two primary models used to project changes in climate in this Assessment were developed at the Canadian Climate Centre and the Hadley Centre in the United Kingdom. They have been peer-reviewed by other scientists and both incorporate similar assumptions about future emissions (both approximate the mid-range emissions scenario described on page 4). These models were the best fit to a list of criteria developed for this Assessment. Climate models developed at the National Center for Atmospheric Research (NCAR), NOAA's Geophysical Fluid Dynamics Laboratory (GFDL), NASA's Goddard Institute for Space Studies (GISS), and Max Planck Institute (MPI) in Germany, were also used in various aspects of the Assessment.

While the physical principles driving these models are similar, the models differ in how they represent the effects of some important processes. Therefore, the two primary models paint different views of 21st century climate. On average over the US, the Hadley model projects a much wetter climate than does the Canadian model, while the Canadian model projects a greater increase in temperature than does the Hadley model. Both projections are plausible, given current understanding. In most climate models, increases in temperature for the US are significantly higher than the global average temperature increase. This is due to the fact that all models project the warming to be greatest at middle to high latitudes, partly because melting snow and ice make the surface less reflective of sunlight, allowing it to absorb more heat. Warming will also be greater over land than over the oceans because it takes longer for the oceans to warm.

On average over the US, the Hadley model projects a wetter climate than does the Canadian model, while the Canadian model projects a greater increase in temperature than does the Hadley model.

Uncertainties about future climate stem from a wide variety of factors, from questions about how to represent clouds and precipitation in climate models to uncertainties about how emissions of greenhouse gases will change. These uncertainties result in differences in climate model projections. Examining these differences aids in understanding the range of risk or opportunity associated with a plausible range of future climate changes. These differences in model projections also raise questions about how to interpret model results, especially at the regional level where projections can differ significantly.

Changes in Temperature over the US Simulated by Climate Models

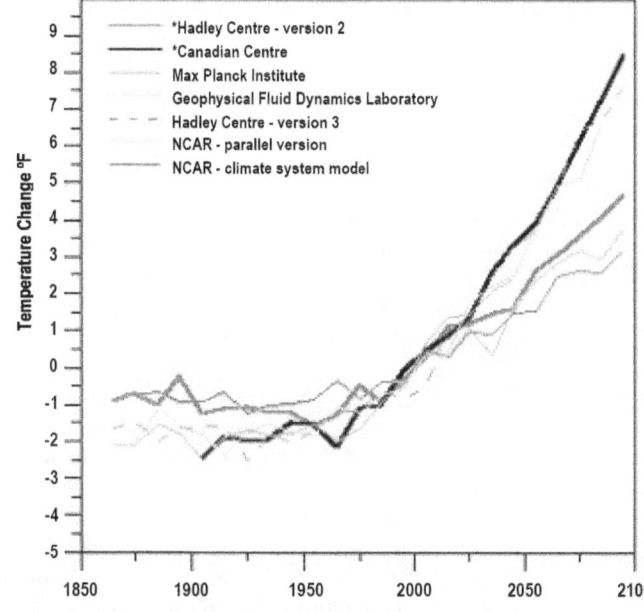

The two primary climate models used in this Assessment have been peer-reviewed and both incorporate similar assumptions about future emissions (both approximate the IPCC "IS92a" scenario with a 1% per year increase in greenhouse gases and growing sulfur emissions).

Range of Projected Warming in the 21st Century

	Global	US
*Hadley Model	+5°F	+5°F
*Canadian Model	+8°F	+9°F
MPI, GFDL and NCAR Models	+3 to 6°F	+3 to 9°F

*The two primary models used in the Assessment.

Simulations from leading climate models of changes in decadal average surface temperature for the conterminous US (excluding Alaska and Hawaii) based on historic and projected changes in atmospheric concentrations of greenhouse gases and sulfate aerosols. The heavy red and black lines indicate the primary models used by the National Assessment. For the 20th century, the models simulate a US temperature rise of about 0.7 to 1.9°F, whereas estimates from observations range from 0.5 to 1.4°F; estimates for the global rise are 0.9 to 1.4°F for models and 0.7 to 1.4°F for observations, suggesting reasonable agreement. For the 21st century, the models project warming ranging from 3 to 6°F for the globe and 3 to 9°F for the US. The two models at the low end of this range assume lower emissions of greenhouse gases than do the other models.

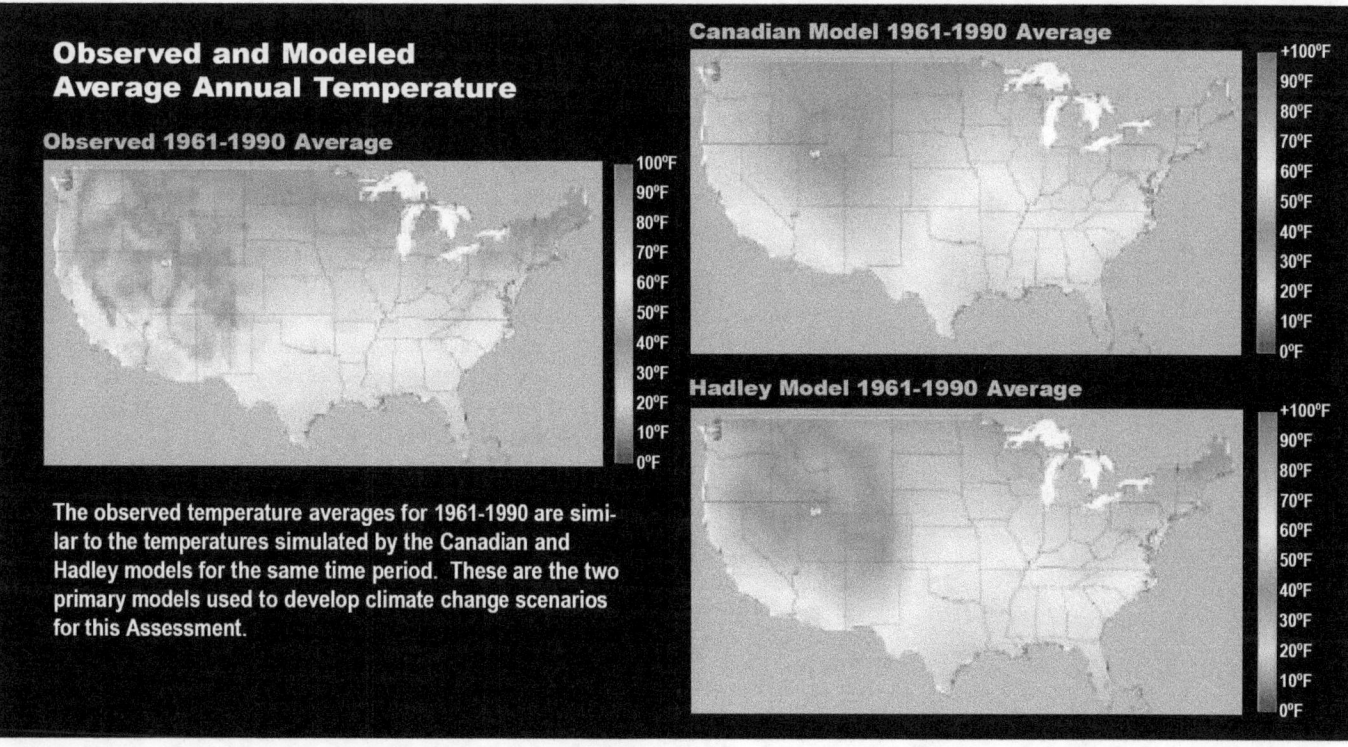

Observed and Modeled Average Annual Temperature

Observed 1961-1990 Average

Canadian Model 1961-1990 Average

Hadley Model 1961-1990 Average

The observed temperature averages for 1961-1990 are similar to the temperatures simulated by the Canadian and Hadley models for the same time period. These are the two primary models used to develop climate change scenarios for this Assessment.

TOOLS FOR ASSESSING CLIMATE CHANGE IMPACTS

Interpreting Climate Scenarios

Model projections of continental-scale and century-long trends are more reliable than projections of shorter-term trends over smaller scales.

Our level of confidence in climate scenarios depends on what aspect is being considered, and over what spatial scale and time period. Increases in greenhouse gases will cause global temperatures to increase. There is less certainty about the magnitude of the increase, because we lack complete knowledge of the climate system and because we do not know how human society and its energy systems will evolve. Similarly, we are confident that higher surface temperatures will cause an increase in evaporation, and hence in precipitation, but less certain about the distribution and magnitude of these changes.

The most certain climate projections are those that pertain to large-scale regions, are given as part of a range of possible outcomes, and are applied to trends over the next century. Model projections of continental-scale and century-long trends are more reliable than projections of shorter-term trends over smaller scales. Projections on a decade-by-decade basis, and projections of transient weather phenomena such as hurricanes, are considerably less certain. Two examples serve to illustrate this point. Most climate models project warming in the eastern Pacific, resulting in conditions that look much like current El Niño conditions. When today's existing El Niño pattern is superimposed on this El Niño-like state, El Niño events would likely be more intense, as would their impacts on US weather. Some recent studies suggest that El Niño and La Niña conditions are likely to become more frequent and intense. Other studies suggest little overall change. While these projections must be interpreted with caution, prudent risk management suggests considering the possibility of increases in El Niño and La Niña intensity and frequency.

The projections are less certain regarding changes in the incidence of tropical storms and hurricanes. Some recent studies suggest that hurricanes will become more intense, while others project little change. It is possible that a 5-10% increase in hurricane wind speed will occur by 2100; confirming this remains an important research issue. Perhaps a more important concern is rainfall during hurricanes. One set of model simulations projects that peak precipitation rates during hurricanes will increase 25-30% by the end of the 21st century. Today, El Niño conditions are associated with increased Pacific and decreased Atlantic hurricane frequencies. La Niña is associated with increased Atlantic hurricane frequencies. However, hurricane formation is dependent on a large number of atmospheric and surface conditions. Given these complex dynamics, projections for changes in the frequency and paths of tropical storms must be viewed with caution.

El Niño and La Niña Effects on the Chance of Landfalling Hurricanes over the 20th Century

El Niño   La Niña

Chance of Occurrence in %

Number of Hurricanes per year

During El Niño and La Niña years, the chance of landfalling hurricanes on the Gulf and Atlantic coasts changes dramatically, as seen in this chart based on data since 1900. During El Niño years the chance of hurricanes is greatly reduced; no more than two hurricanes have ever made landfall during an El Niño year. On the other hand, during La Niña years, the chance of hurricanes greatly increases; there has been nearly a 40% chance of three or more hurricanes making landfall during a La Niña year.

A Continually Changing Climate and the Potential for Surprises

I t is essential to note that the 21st century's climate,unlike that of the preceeding thousand years,is not expected to be stable but is very likely to be in a constant state of change. For example,the duration and amount of ice in the Great Lakes is expected to decrease. It is possible that in the short term an increase in "lake effect" snows would be a consequence during mid-winter, though they would likely decrease in the long term. Across the nation,as climate continues to warm,precipitation is very likely to increasingly fall as rain rather than snow. Such continuously changing climate presents a special challenge for human adaptation.

In addition,there is the potential for "surprises." Because climate is highly complex,it is important to remember that it might surprise us with sudden or discontinuous change,or by otherwise evolving quite differently from what is expected. Surprises challenge humans' ability to adapt,because of how quickly and unexpectedly they occur. For example,what if the Pacific Ocean warms in such a way that El Niño events become much more extreme? This could reduce the frequency, but perhaps not the strength,of hurricanes along the East Coast,while on the West Coast,more severe winter storms, extreme precipitation events,and damaging winds could become common. What if large quantities of methane,a potent greenhouse gas currently frozen in icy Arctic tundra and sediments,began to be released to the atmosphere by warming,potentially creating an amplifying "feedback loop" that would cause even more warming? We simply do not know how far the climate system or other systems it affects can be pushed before they respond in unexpected ways.

There are many examples of potential surprises,each of which would have large consequences. Most of these potential outcomes are rarely reported,in this study or elsewhere. Even if the chance of any particular surprise happening is small,the chance that at least one such surprise will occur is much greater. In other words,while we can't know which of these events will occur, it is likely that one or more will eventually occur.

Another caveat is appropriate: climate scenarios are based on emissions scenarios for various gases. The development of new energy technologies,the speed of population growth, and changes in consumption rates each have the potential to alter these emissions in the future,and hence the rate of climate change.

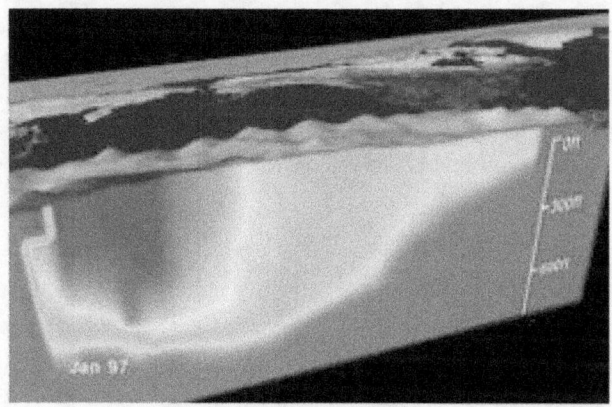

Water temperature profile in the Pacific Ocean, January 1997.

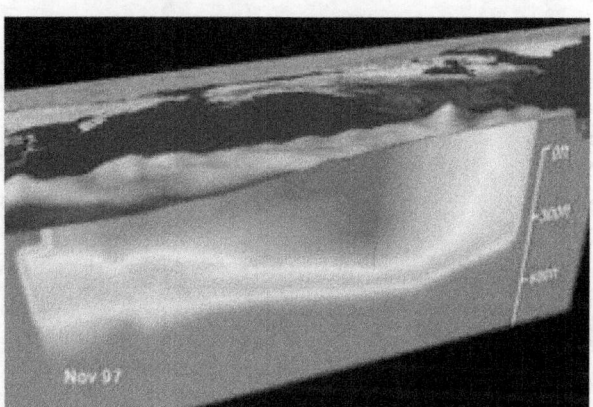

Water temperature profile in the Pacific Ocean, November 1997.

During El Niño conditions, the equatorial pool of warm water (shown in red) expands and moves eastward to span the entire equatorial Pacific east of the dateline. This dramatic warming affects global atmospheric circulation including effects on the jet stream, winter storms, and tropical storms.

LOOKING AT AMERICA'S CLIMATE

Past and Future US Temperature Change

Observations from 1200 weather stations across the US show that temperatures have increased over the past century, on average by almost 1°F (0.6°C). The coastal Northeast, the upper Midwest, the Southwest, and parts of Alaska have experienced increases in the annual average temperature approaching 4°F (2°C) over the past 100 years. The rest of the nation has experienced less warming. The Southeast and southern Great Plains have actually experienced a slight cooling over the 20th century, but since the 1970s have had increasing temperatures as well. The largest observed warming across the nation has occurred in winter.

Average warming in the US is projected to be somewhat greater than for the world as a whole over the 21st century. In the Canadian model scenario, increases in annual average temperature of 10°F (5.5°C) by the year 2100 occur across the central US with changes about half this large along the east and west coasts. Seasonal patterns indicate that projected changes will be particularly large in winter, especially at night. Large increases in temperature are projected over much of the South in summer, dramatically raising the heat index (a measure of discomfort based on temperature and humidity).

In the Hadley model scenario, the eastern US has temperature increases of 3-5°F (2-3°C) by 2100 while the rest of the nation warms more, up to 7°F (4°C), depending on the region.

In both models, Alaska is projected to experience more intense warming than the lower 48, and in fact, this warming is already well underway. In contrast, Hawaii and the Caribbean islands are likely to experience less warming than the continental US, because they are at lower latitudes and are surrounded by ocean, which warms more slowly than land.

Both the Canadian and Hadley model scenarios project substantial warming during the 21st century. The warming is considerably greater in the Canadian model, with most of the continental US experiencing increases from 5 to 15°F. In this model, the least warming occurs in the West and along the Atlantic and Gulf Coasts. In the Hadley model, annual temperatures are projected to increase from 3 to 7°F, with the largest warming occurring in the western half of the country.

Temperature Change

How to read these maps: The color scale indicates changes in temperature in °F over a 100 year period. For example, at 0°F there is no change; at +10°F there is a 10°F increase from the begining to the end of the century.

Observed 20th Century

The change in the annual average temperature over the 20th century has a distinctive pattern. Most of the US has warmed, in some areas by as much as 4°F. Only portions of the southeastern US have experienced cooling, and this was primarily due to the cool decades of the 1960s and 1970s. Temperatures since then have reached some of the highest levels of the century.

Canadian Model 21st Century

Hadley Model 21st Century

Minimum Temperature in the US (annual average)

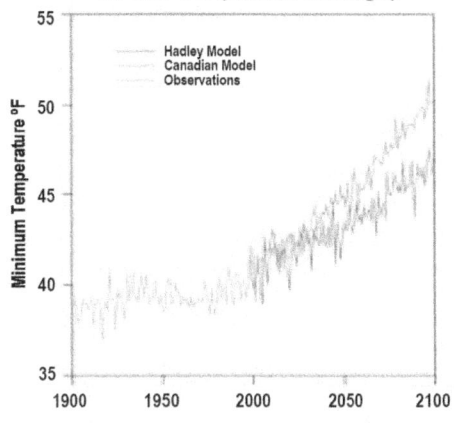

Maximum Temperature in the US (annual average)

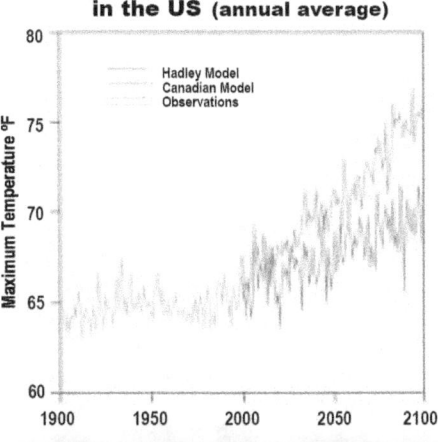

Average US warming is projected to be somewhat greater than global average warming over the 21st century. Large increases in temperature are projected over much of the South in summer, dramatically raising the heat index (a measure of discomfort based on temperature and humidity).

The annual average of minimum and maximum temperatures are compiled from the daily lows and highs. These graphs show the lows and highs, averaged over the year and over the lower 48 states. The green line shows observed temperatues while the red and blue lines are model projections for the future.

The minimum and maximum temperatures are important because, far more than the average, they influence such things as human comfort, heat and cold stress in plants and animals, maintenance of snow-pack, and pest populations (many pests are killed by low temperatures; a rise in the minimum often allows more pests to survive).

July Heat Index Change

The projected changes in the heat index for the Southeast are the most dramatic in the nation with the Hadley model suggesting increases of 8 to 15°F for the southernmost states, while the Canadian model projects increases above 25°F for much of the region.

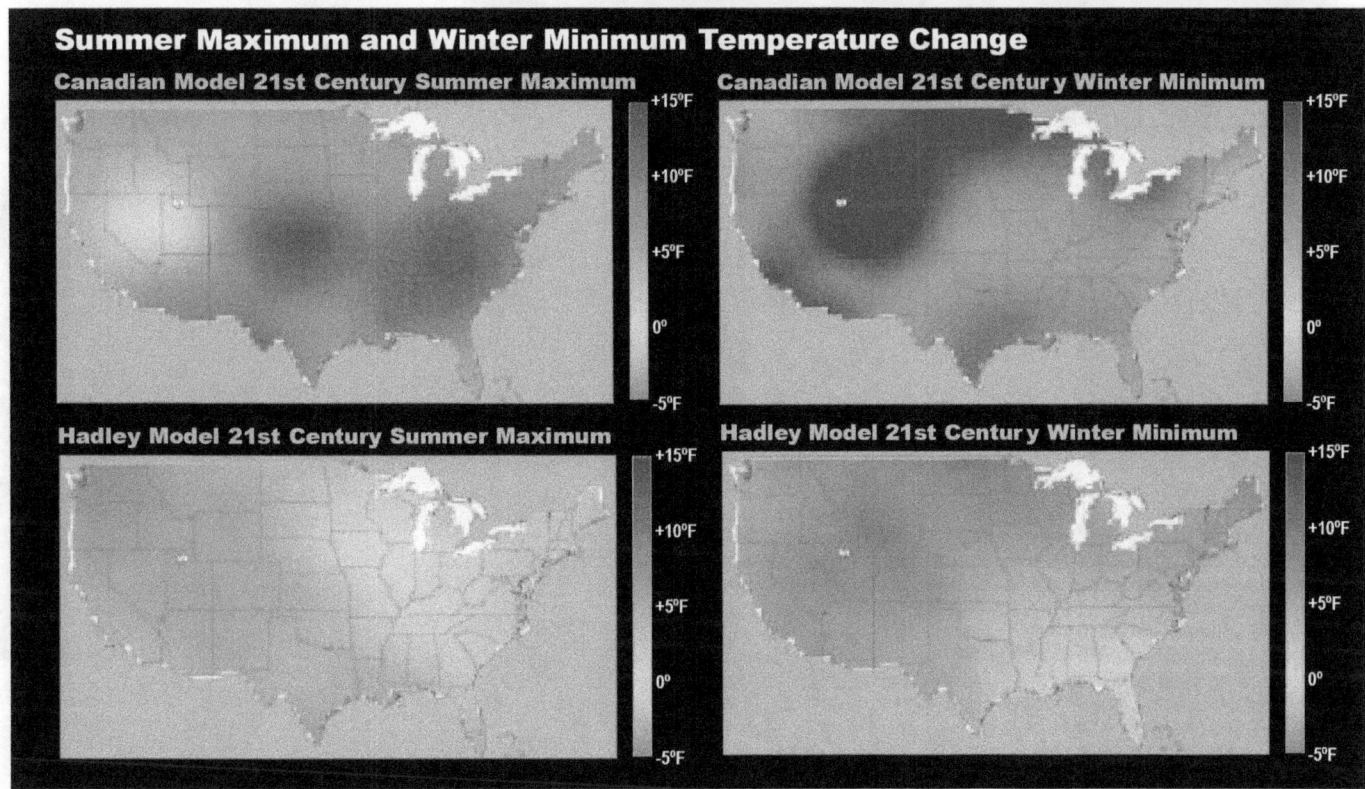

Summer Maximum and Winter Minimum Temperature Change

LOOKING AT AMERICA'S CLIMATE

Changes in Precipitation

Average US precipitation has increased by 5-10% over the last century with much of that due to an increase in the frequency and intensity of heavy rainfall. Precipitation increases have been especially noteworthy in the Midwest, southern Great Plains, and parts of the West and Pacific Northwest. Decreases have been observed in the northern Great Plains.

For the 21st century, the Canadian model projects that percentage increases in precipitation will be largest in the Southwest and California, while east of the Rocky Mountains, the southern half of the nation is projected to experience a decrease in precipitation. The percentage decreases are projected to be particularly large in eastern Colorado and western Kansas, and across an arc running from Louisiana to Virginia. Projected decreases in precipitation are most evident in the Great Plains during summer and in the East during both winter and summer. The increases in precipitation projected to occur in the West, and the smaller increases in the Northwest, are projected to occur mainly in winter.

In the Hadley model, the largest percentage increases in precipitation are projected to be in the Southwest and Southern California, but the increases are smaller than those projected by the Canadian model. In the Hadley model, the entire US is projected to have increases in precipitation, with the exception of small areas along the Gulf Coast and in the Pacific Northwest. Precipitation is projected to increase in the eastern half of the nation and in southern California and parts of Nevada and Arizona in summer, and in every region during the winter, except the Gulf States and northern Washington and Idaho.

In both the Hadley and Canadian models, most regions are projected to experience an increase in the frequency of heavy precipitation events. This is especially notable in the Hadley model, but the Canadian model shows the same characteristic.

While the actual amounts are modest, the large percentage increases in rainfall projected for the Southwest are related to increases in atmospheric moisture and storm paths. A warmer Pacific would pump moisture into the region and there would also be a southward shift in Pacific Coast storm activity. In the Sierra Nevada and Rocky Mountains, much of the increased precipitation is likely to fall as rain rather than snow, causing a reduction in mountain snow packs.

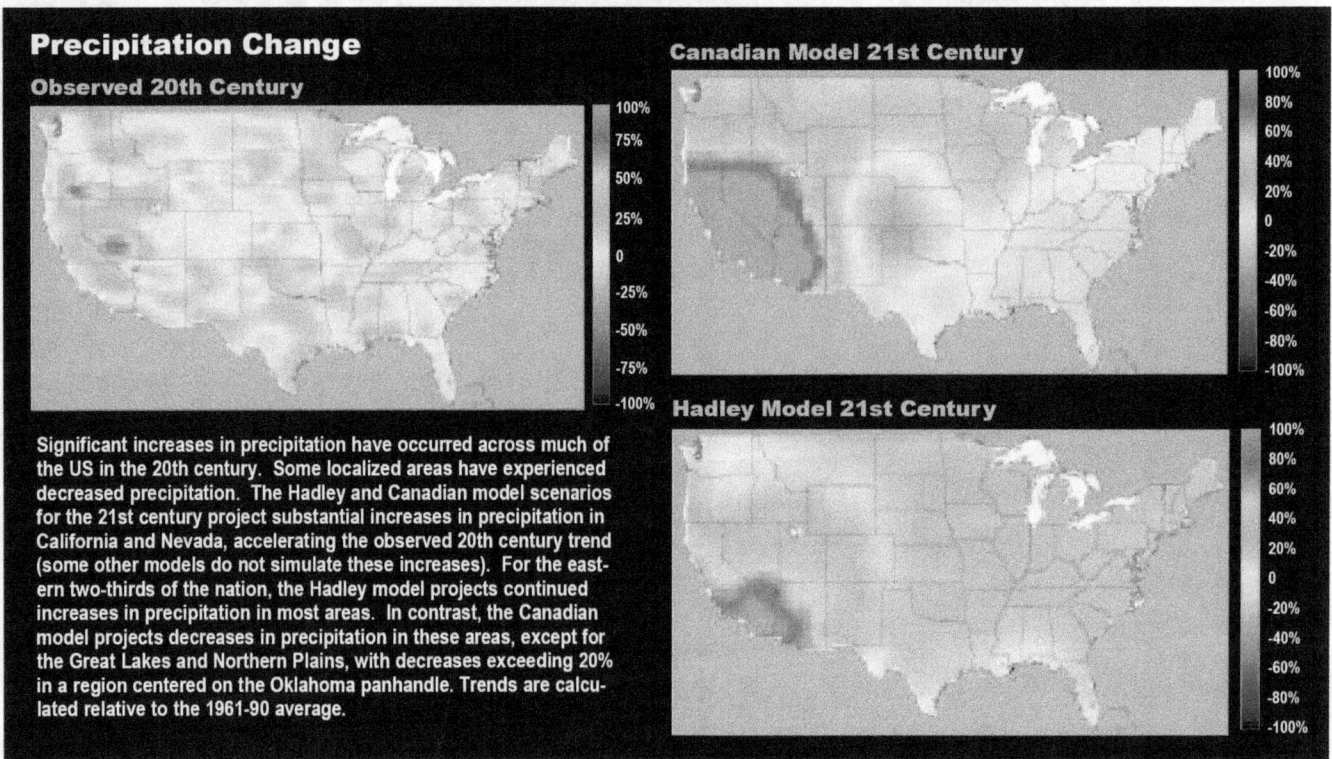

Precipitation Change
Observed 20th Century

Canadian Model 21st Century

Hadley Model 21st Century

Significant increases in precipitation have occurred across much of the US in the 20th century. Some localized areas have experienced decreased precipitation. The Hadley and Canadian model scenarios for the 21st century project substantial increases in precipitation in California and Nevada, accelerating the observed 20th century trend (some other models do not simulate these increases). For the eastern two-thirds of the nation, the Hadley model projects continued increases in precipitation in most areas. In contrast, the Canadian model projects decreases in precipitation in these areas, except for the Great Lakes and Northern Plains, with decreases exceeding 20% in a region centered on the Oklahoma panhandle. Trends are calculated relative to the 1961-90 average.

This would tend to increase wintertime river flows and decrease summertime flows in the West. Across the Northwest, and the central and eastern US, the two model projections of precipitation change are in less agreement. These differences will be resolved only by improvements in climate modeling.

Changes in Soil Moisture

Soil moisture is critical for both agriculture and natural ecosystems. Soil moisture levels are determined by an intricate interplay among precipitation, evaporation, run-off, and soil drainage. By itself, an increase in precipitation would increase soil moisture. However, higher air temperatures will increase the rate of evaporation and, in some areas, remove moisture from the soil faster than it can be added by precipitation. Under these conditions, some regions are likely to become drier even though their rainfall increases.

In fact, soil moisture has already decreased in portions of the Great Plains and Eastern Seaboard, where precipitation has increased but air temperature has risen.

Since soil moisture projections reflect both changes in precipitation and in evaporation associated with warming, the differences between the two models are accentuated in the soil moisture projections. For example, in the Canadian model, soil moisture decreases of more than 50% are common in the Central Plains due to the combination of precipitation reductions exceeding 20% and temperature increases exceeding 10°F. In the Hadley model, this same region experiences more modest warming of about 5°F and precipitation increases of around 20%, generally resulting in soil moisture increases.

Increased drought becomes a national problem in the Canadian model. Intense drought tendencies occur in

the region east of the Rocky Mountains and throughout the Mid-Atlantic-Southeastern states corridor. Increased tendencies toward drought are also projected in the Hadley model for regions immediately east of the Rockies. California and Arizona, plus a region from eastern Nebraska to Virginia's coastal plain, experience decreases in drought tendency. The differences in soil moisture and drought tendencies will be significant for water supply, agriculture, forests, and lake levels.

In both the Hadley and Canadian models, most regions are projected to see an increase in the frequency of heavy precipitation events.

Higher air temperatures will increase the rate of evaporation and, in some areas, remove moisture from the soil faster than it can be added by precipitation.

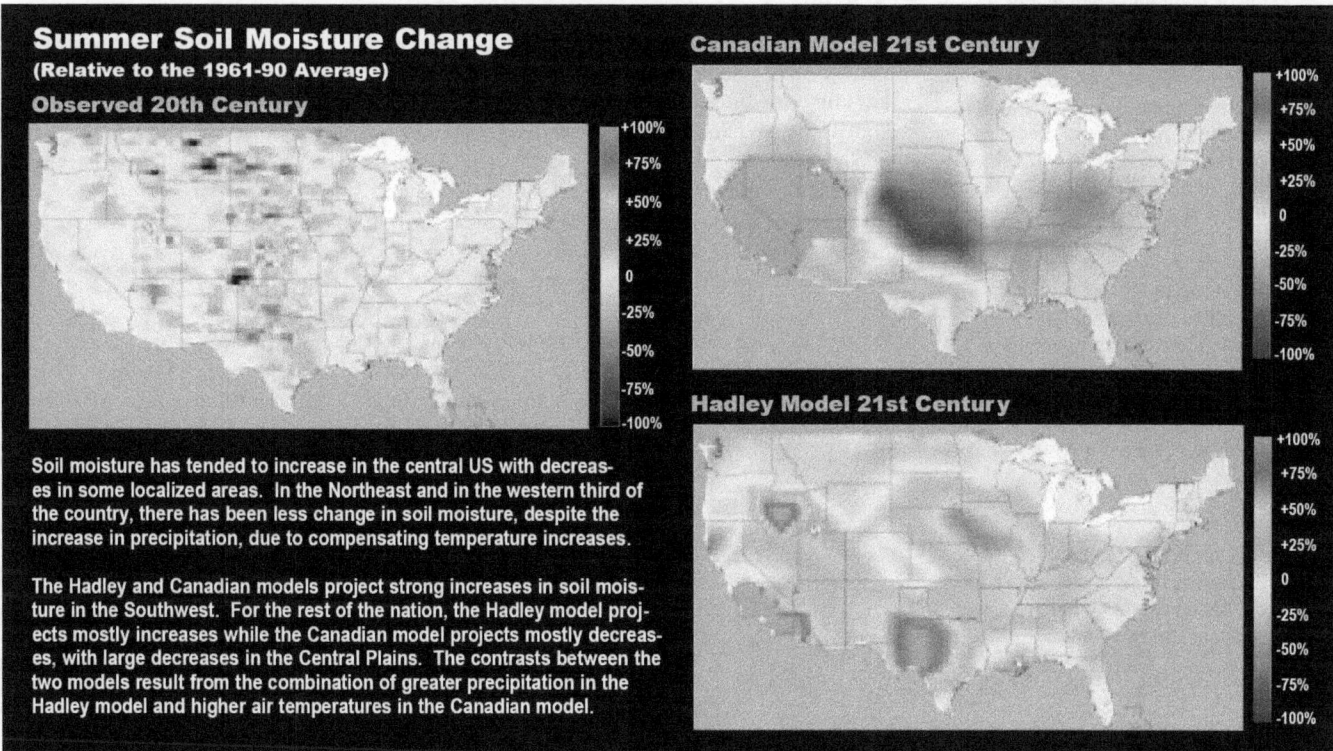

Summer Soil Moisture Change
(Relative to the 1961-90 Average)

Observed 20th Century

Canadian Model 21st Century

Hadley Model 21st Century

Soil moisture has tended to increase in the central US with decreases in some localized areas. In the Northeast and in the western third of the country, there has been less change in soil moisture, despite the increase in precipitation, due to compensating temperature increases.

The Hadley and Canadian models project strong increases in soil moisture in the Southwest. For the rest of the nation, the Hadley model projects mostly increases while the Canadian model projects mostly decreases, with large decreases in the Central Plains. The contrasts between the two models result from the combination of greater precipitation in the Hadley model and higher air temperatures in the Canadian model.

ECOSYSTEMS IN THE FUTURE

The natural vegetation covering about 70% of the US land surface is strongly influenced both by the climate and by the atmospheric carbon dioxide (CO_2) concentration. To provide a common base of information about potential changes in vegetation across the nation for use in the regional and sector studies, specialized ecosystem models were run using the two major climate model scenarios selected for this Assessment. A summary of the national level results follows. Agricultural and production forestry systems are the focus of separate sections of this Overview report.

What are Ecosystems?

Ecosystems are communities of plants, animals, microbes, and the physical environment in which they exist. They can be characterized by their biological richness, by the magnitude of flows of energy and materials between their constituent species and their physical environment, and by the interactions among the biological species themselves, that is, by which species are predators and prey, which are competitors, and which are symbiotic.

Ecologists often categorize ecosystems by their dominant vegetation – the deciduous broad-leafed forest ecosystems of New England, the short-grass prairie ecosystems of the Great Plains, the desert ecosystems of the Southwest. The term "ecosystem" is used not only to describe natural systems (such as coral reefs, alpine meadows, old growth forests, or riparian habitats), but also for plantation forests and agricultural systems, although these ecosystems obviously differ in many important ways from the natural ecosystems they have replaced.

Ecosystems Supply Vital Goods and Services

While we value natural ecosystems in their own right, ecosystems of all types, from the most natural to the most extensively managed, produce a variety of goods and services that benefit humans. Some of these enter the market and contribute directly to the economy. Thus, forests as sources of timber and pulpwood, and agro-ecosystems as sources of food are important to us. But ecosystems also provide a set of un-priced services that are valuable, but that typically are not traded in the marketplace. There is no current market, for example, for the services that forests and wetlands provide for improving water quality, regulating stream flow, and providing some measure of protection from floods. However, these services are very valuable to society.

Ecosystems are also valued for recreational, aesthetic, and ethical reasons. These are also difficult to value monetarily, but are nevertheless important. The bird life of the coastal marshes of the Southeast and the brilliant autumn colors of the New England forests are treasured components of our regional heritages, and important elements of our quality of life.

What to Expect with Climate Change

• Changes in the productivity and carbon storage capacity of ecosystems, decreases in some places and increases in others, are very likely.

• Shifts in the distribution of major plant and animal species are likely.

• Some ecosystems such as alpine meadows are likely to disappear in some places because the new local climate will not support them or there are barriers to their movement.

• In many places, it is very likely that ecosystem services, such as air and water purification, landscape stabilization against erosion, and carbon storage capacity will be reduced. These losses will likely occur in the wake of episodic, large-scale disturbances that trigger species migrations or local extinctions.

• In some places, it is very likely that ecosystems services will be enhanced where climate-related stresses are reduced.

Climate and Ecosystems

Climatic conditions determine where individual species of plants and animals can live, grow, and reproduce. Thus, the collections of species that we are familiar with – the southeastern mixed deciduous forest, the desert ecosystems of the arid Southwest, or the productive grasslands of the Great Plains – are influenced by climate as well as other factors such as land-use. The species in some ecosystems are so strongly influenced by the climate to which they are adapted that they are vulnerable even to modest climate changes. For example, alpine meadows at high elevations in the West exist where they do entirely because the plants that comprise them are adapted to the cold conditions that would be too harsh for other species in the region. The desert vegetation of the Southwest is adapted to the high summer temperatures and aridity of the region. Forests in the east are adapted to relatively high rainfall and soil moisture; if drought conditions were to persist, grasses and shrubs could begin to out-compete tree seedlings, leading to completely different ecosystems.

There are also many freshwater and marine examples of sensitivities to climate variability and change. In aquatic ecosystems, for example, many fish can breed only in water that falls within a narrow range of temperatures. Thus, species of fish that are adapted to cool waters can quickly become unable to breed successfully if water temperatures rise. Wetland plant species can adjust to rising sea levels by dispersing to new locations, within limits. Too rapid sea-level rise can surpass the ability of the plants to disperse, making it impossible for coastal wetland ecosystems to re-establish themselves.

The species in some ecosystems are so strongly influenced by the climate to which they are adapted that they are vulnerable even to modest climate changes.

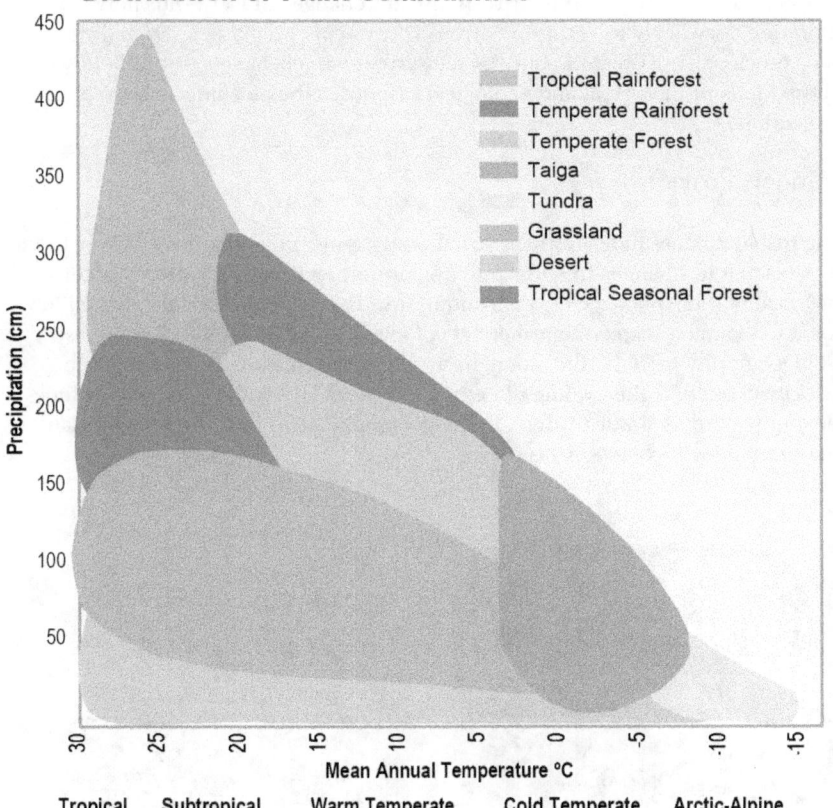

Distribution of Plant Communities

- Tropical Rainforest
- Temperate Rainforest
- Temperate Forest
- Taiga
- Tundra
- Grassland
- Desert
- Tropical Seasonal Forest

Precipitation (cm) — axis values: 50, 100, 150, 200, 250, 300, 350, 400, 450

Mean Annual Temperature °C — axis values: 30, 25, 20, 15, 10, 5, 0, -5, -10, -15

Tropical   Subtropical   Warm Temperate   Cold Temperate   Arctic-Alpine

Both temperature and precipitation limit the distribution of plant communities. The climate (temperature and precipitation) zones of some of the major plant communities (such as temperate forests, grasslands, and deserts) in the US are shown in this figure. Note that grasslands' zone encloses a wide range of environments. This zone can include a mixture of woody plants with the grasses. The shrublands and woodlands of the West are examples of grass/woody vegetation mixes that occur in the zone designated as grasslands.

With climate change, the areas occupied by these zones will shift relative to their current distribution. Plant species are expected to shift with their climate zones. The new plant communities that result from these shifts are likely to be different from current plant communities because individual species will very likely migrate at different rates and have different degrees of success in establishing themselves in new places.

ECOSYSTEMS IN THE FUTURE

Effects of Increased CO_2 Concentration on Plants

The ecosystem models used in this Assessment consider not only changes in climate, but also increases in atmospheric CO_2. The atmospheric concentration of CO_2 affects plant species in ecosystems since it has a direct physiological effect on photosynthesis, the process by which plants use CO_2 to create new biological material. Higher concentrations of CO_2 generally enhance plant growth if the plants also have sufficient water and nutrients, such as nitrogen, to sustain this enhanced growth. For this reason, the CO_2 levels in commercial greenhouses are sometimes boosted in order to stimulate plant growth. In addition, higher CO_2 levels can raise the efficiency with which plants use water. Different types of plants respond at different rates to increases in atmospheric CO_2, resulting in a divergence of growth rates due to CO_2 increase. Some species grow faster, but provide reduced nutritional value. The effects of increased CO_2 level off at some point; thus, continuing to increase CO_2 levels will not result in increased plant growth indefinitely. There is still much we do not understand about the CO_2 "fertilization" effect, its limits, and its direct and indirect implications.

Species Responses to Changes in Climate and CO_2

The responses of ecosystems to changes in climate and CO_2 are made up of the individual responses of their constituent species and how they interact with each other. Species in current ecosystems can differ substantially in their tolerances of changes in temperature and precipitation, and in their responses to changes in CO_2; thus, new climate conditions are very likely to result in current ecosystems breaking apart, and new assemblages of species being created. Current ecosystem models have great difficulty in predicting these kinds of biological and ecological responses, thus leading to large uncertainties in projections.

What the Models Project

Modeling results to date indicate that natural ecosystems on land are very likely to be highly sensitive to changes in surface temperature, precipitation patterns, other climate parameters, and atmospheric CO_2 concentrations. Two types of models utilized in this Assessment to examine the ecological effects of climate change are biogeochemistry models and biogeography models. Biogeochemistry models simulate changes in basic ecosystem processes such as the cycling of carbon, nutrients, and water (ecosystem function). Biogeography models simulate shifts in the geographic distribution of major plant species and communities (ecosystem structure).

Species in current ecosystems can differ substantially in their tolerances of changes in temperature and precipitation, and in their responses to changes in CO_2; thus, new climate conditions are very likely to result in current ecosystems breaking apart, and new assemblages of species being created.

The biogeochemistry models used in this analysis generally simulate increases in the amount of carbon in vegetation and soils over the next 30 years for the continental US as a whole. These probable increases are small – in the range of 10% or less, and are not uniform across the country. In fact, for some regions the models simulate carbon losses over the next 30 years. One of the biogeochemistry models, when operating with the Canadian climate scenario, simulates that by about 2030, parts of the Southeast will likely lose up to 20% of the carbon from their forests. A carbon loss by a forest is treated as an indication that it is in decline. The same biogeochemistry model, when operating with the Hadley climate scenario, simulates that forests in the same part of the Southeast will likely gain between 5 and 10% in carbon in trees over the next 30 years.

Why do the two climate scenarios result in opposite ecosystem responses in the Southeast? The Canadian climate scenario shows the Southeast as a hotter and drier place in the early decades of the 21st century than does the Hadley scenario. With the Canadian scenario, forests will be under stress due to insufficient moisture, which causes them to lose more carbon in respiration than they gain in photosynthesis. In contrast, the Hadley scenario simulates relatively plentiful soil moisture, robust tree growth, and forests that accumulate carbon.

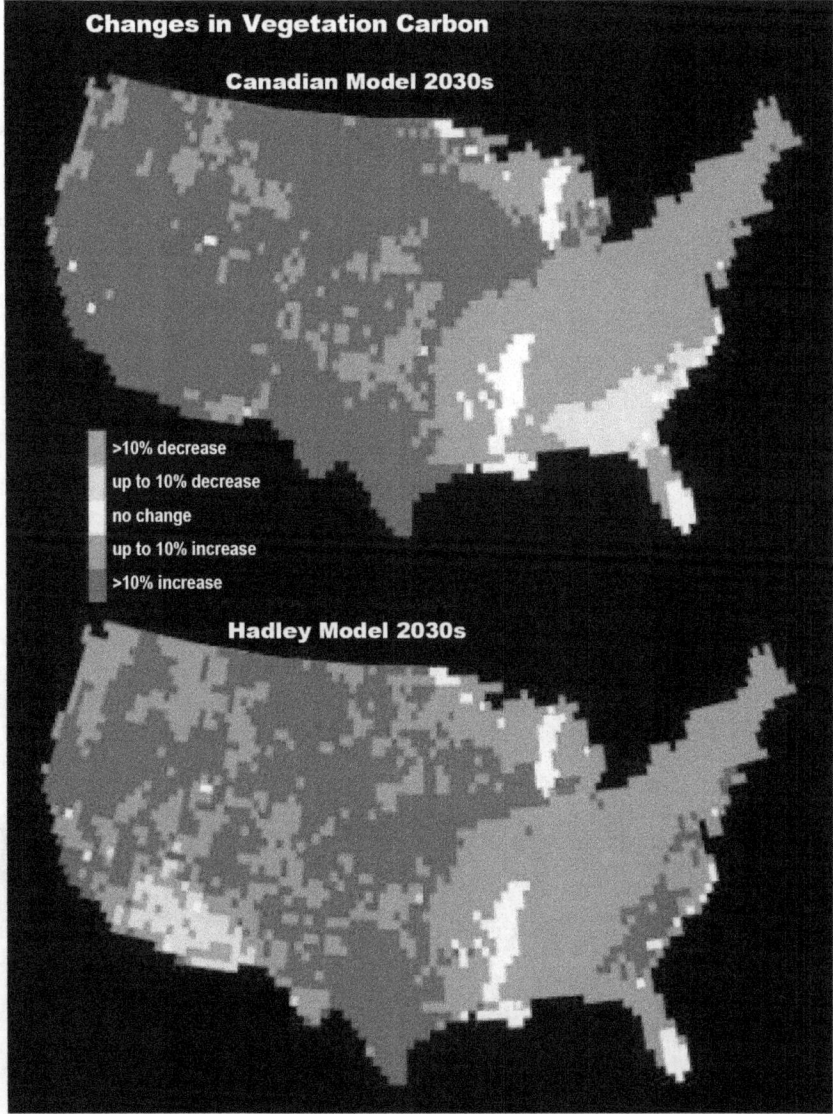

Changes in Vegetation Carbon

Canadian Model 2030s

> >10% decrease
> up to 10% decrease
> no change
> up to 10% increase
> >10% increase

Hadley Model 2030s

The maps above show projections of relative changes in vegetation carbon between 1990 and the 2030s for two climate scenarios. Under the Canadian model scenario, vegetation carbon losses of up to 20% are projected in some forested areas of the Southeast in response to warming and drying of the region by the 2030s. A carbon loss by forests is treated as an indication that they are in decline. Under the same scenario, vegetation carbon increases of up to 20% are projected in the forested areas in the West that receive substantial increases in precipitation. Output from TEM (Terrestrial Ecosystem Model) as part of the VEMAP II (Vegetation Ecosystem Modeling and Analysis Project) study.

ECOSYSTEMS IN THE FUTURE

Will disturbances caused by climate change be regular and small or will they be episodic and large? The latter category of disturbances is likely to have a negative impact on ecosystem services; the ability of ecosystems to cleanse the air and water, stabilize landscapes against erosion, and store carbon, for example, are very likely to be diminished.

Prolonged stress due to insufficient soil moisture can make trees more susceptible to insect attack, lead to plant death, and increase the probability of fire as dead plant material adds to an ecosystem's "fuel load." The biogeography models used in this analysis simulate at least part of this sequence of climate-triggered events in ecosystems as a prelude to shifts in the geographic distribution of major plant species. One of the biogeography models, when operating with the Canadian climate scenario, simulates that towards the end of the 21st century, a hot dry climate in the Southeast will result in the replacement of the current mixed evergreen and deciduous forests by savanna/woodlands and grasslands, with much of the change involving fire. This change in habitat type in the Southeast would imply that the animal populations of the region would also change, although the biogeography models are not designed to simulate these changes. The same biogeography model, when operating with the Hadley scenario, simulates a slight northward expansion of the mixed evergreen and deciduous forests of the Southeast with no significant contraction along the southern boundary. Other biogeography models show similar results.

Major Uncertainties

Major uncertainties exist in the biogeochemistry and biogeography models. For example, ecologists are uncertain about how increases in atmospheric CO_2 affect the carbon and water cycles in ecosystems. What they assume about these CO_2 effects can significantly influence model simulation results. One of these models was used to show the importance of testing these assumptions. Consideration of climate change alone

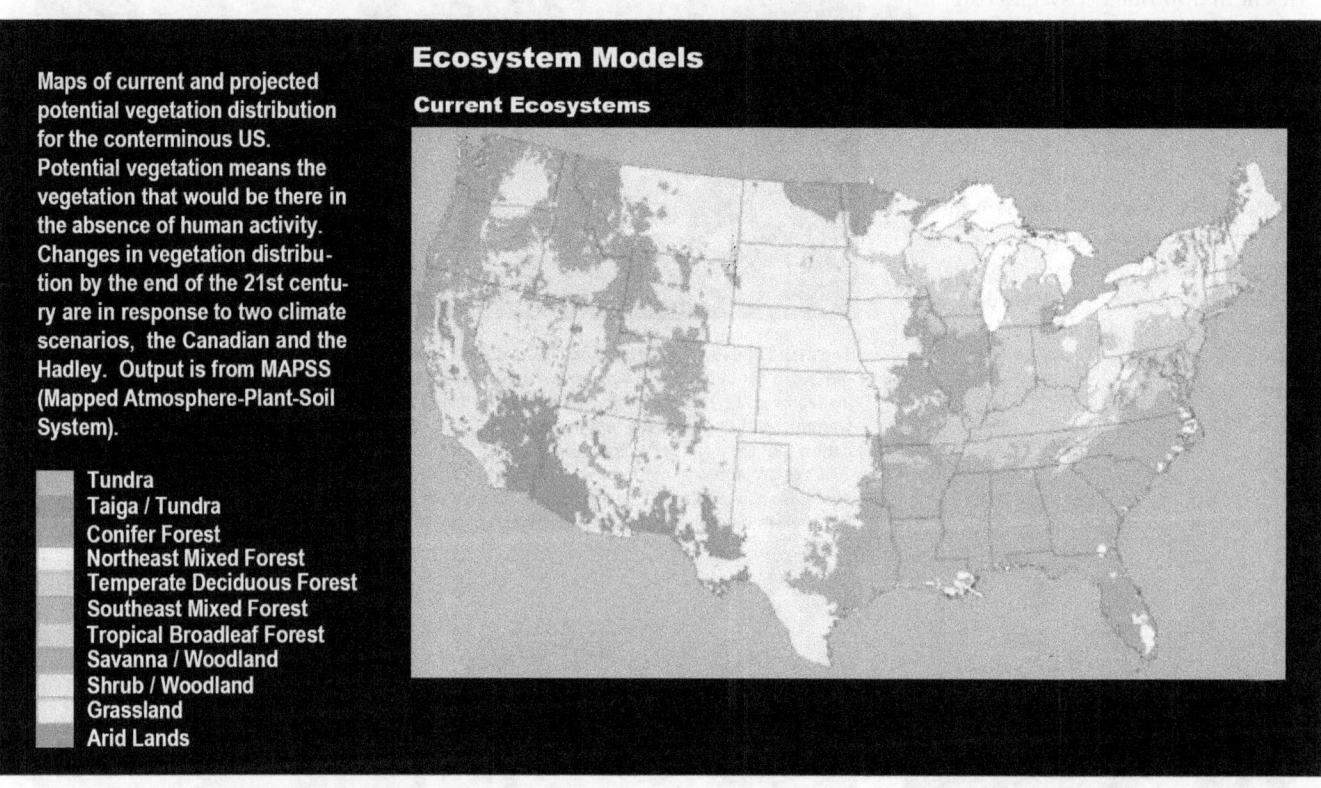

Maps of current and projected potential vegetation distribution for the conterminous US. Potential vegetation means the vegetation that would be there in the absence of human activity. Changes in vegetation distribution by the end of the 21st century are in response to two climate scenarios, the Canadian and the Hadley. Output is from MAPSS (Mapped Atmosphere-Plant-Soil System).

Ecosystem Models

Current Ecosystems

- Tundra
- Taiga / Tundra
- Conifer Forest
- Northeast Mixed Forest
- Temperate Deciduous Forest
- Southeast Mixed Forest
- Tropical Broadleaf Forest
- Savanna / Woodland
- Shrub / Woodland
- Grassland
- Arid Lands

results in a 10% decrease in plant productivity. Consideration of both climate and CO_2 effects results in an increase in plant productivity of 10%. This illustrates the importance of resolving uncertainties about the effects of CO_2 on ecosystems.

With respect to biogeography models, scientists are uncertain about the frequency and size of disturbances produced by factors such as fire and pests that initiate changes in the distribution of major plant and animal species. Will disturbances caused by climate change be regular and small or will they be episodic and large? The latter category of disturbances is likely to have a negative impact on ecosystems services; the ability of ecosystems to cleanse the air and water, stabilize landscapes against erosion, and store carbon, for example, are very likely to be diminished.

Ecologists are uncertain about how increases in atmospheric CO_2 affect the carbon and water cycles in ecosystems. What they assume about these CO_2 effects can significantly influence model simulation results.

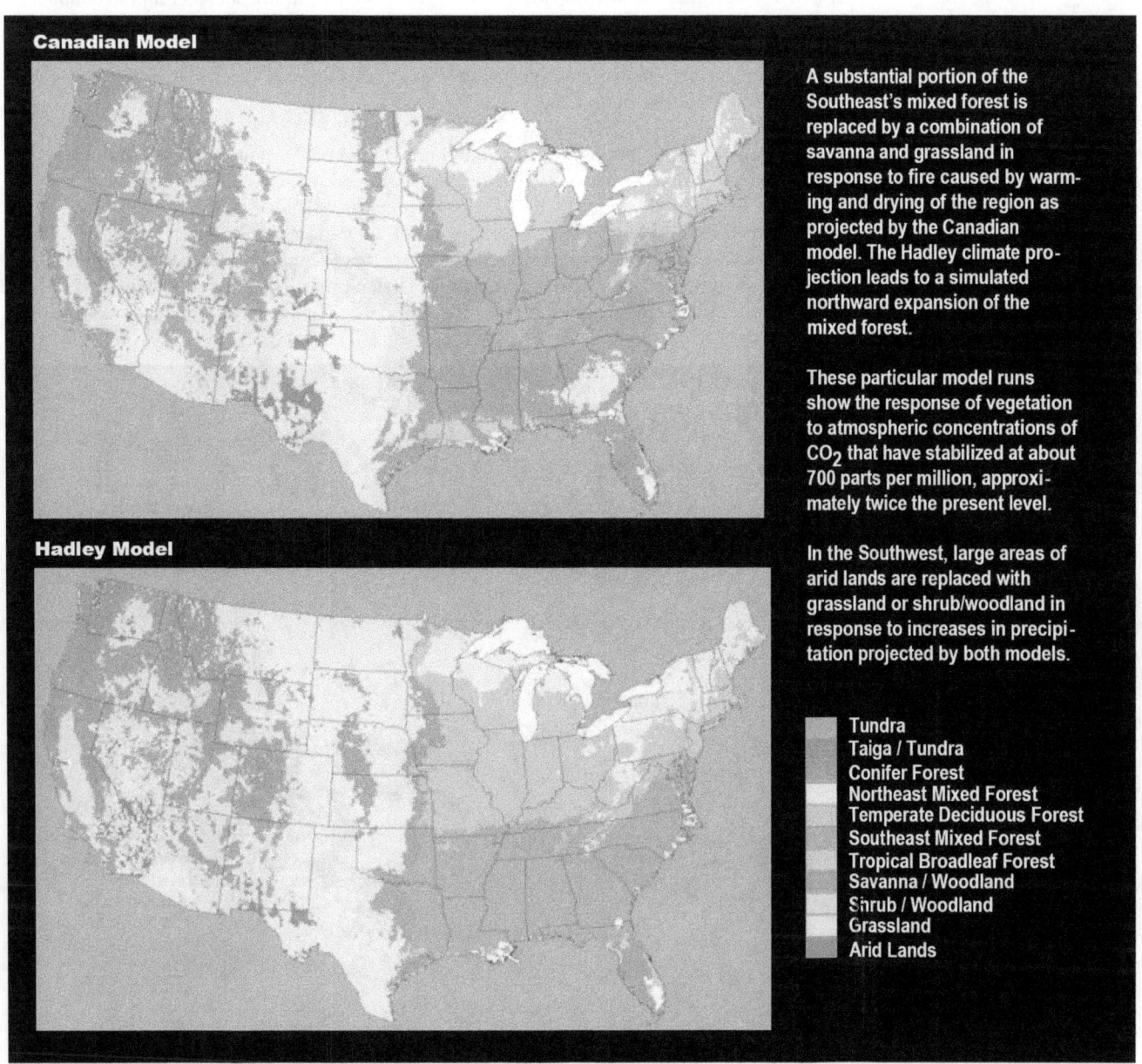

Canadian Model

Hadley Model

A substantial portion of the Southeast's mixed forest is replaced by a combination of savanna and grassland in response to fire caused by warming and drying of the region as projected by the Canadian model. The Hadley climate projection leads to a simulated northward expansion of the mixed forest.

These particular model runs show the response of vegetation to atmospheric concentrations of CO_2 that have stabilized at about 700 parts per million, approximately twice the present level.

In the Southwest, large areas of arid lands are replaced with grassland or shrub/woodland in response to increases in precipitation projected by both models.

Tundra
Taiga / Tundra
Conifer Forest
Northeast Mixed Forest
Temperate Deciduous Forest
Southeast Mixed Forest
Tropical Broadleaf Forest
Savanna / Woodland
Shrub / Woodland
Grassland
Arid Lands

OUR CHANGING NATION

Over the 21st century, assuming no major wars or other catastrophes, growth is likely to continue. However, the specifics of future US growth depend upon many uncertain factors such as technological change, world trade, market conditions, and immigration.

C limate variability and change do not occur in isolation, but in an evolving, dynamic social and economic context. This context is very likely to affect the character and magnitude of climate impacts. Socioeconomic conditions are important drivers of climate change, and also influence the way society responds to change. The prosperity and structure of the economy, the technologies available and in use, and the settlement patterns and demographic structure of the population, are all very likely to contribute to how and how much climate change will matter to Americans, and what they can and might wish to do about it.

Thinking explicitly about socioeconomic futures is speculative, but doing a coherent assessment of future climate impacts requires that potential future socioeconomic conditions be considered. Failing to explicitly consider these conditions risks making the assumption that the future will be largely like the present – an assumption that is virtually certain to be wrong. To see how wrong, one need only compare America's society and economy today to that of 100, 50, or even 25 years ago.

To guide our thinking about socioeconomic futures, this Assessment developed three illustrative socioeconomic scenarios, which project high, medium, and low growth trends for the US population and economy through the 21st century. These scenarios necessarily involve uncertainties that grow large by the end of the century, as the figures show. Nevertheless, they represent a plausible range of socioeconomic conditions that could affect climate impacts and response capabilities. Using multiple scenarios avoids the errors of attempting specific predictions, or assuming no change at all. Region and sector

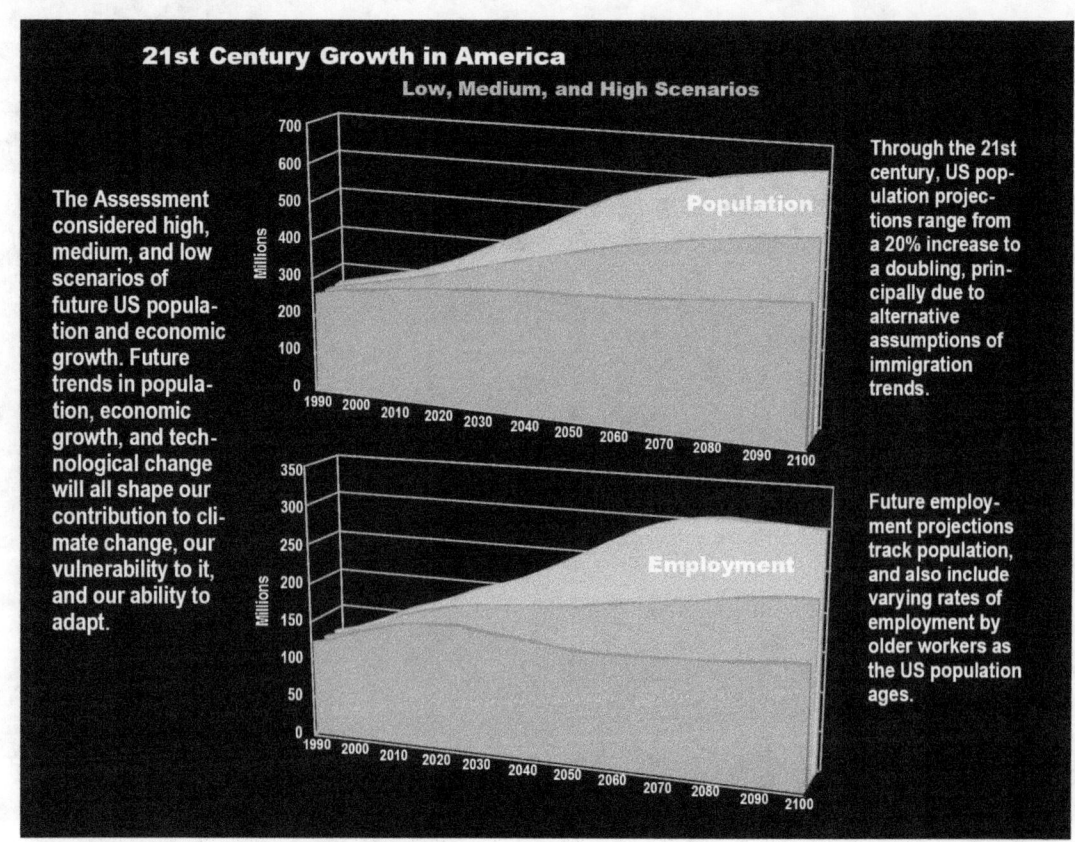

21st Century Growth in America
Low, Medium, and High Scenarios

The Assessment considered high, medium, and low scenarios of future US population and economic growth. Future trends in population, economic growth, and technological change will all shape our contribution to climate change, our vulnerability to it, and our ability to adapt.

Through the 21st century, US population projections range from a 20% increase to a doubling, principally due to alternative assumptions of immigration trends.

Future employment projections track population, and also include varying rates of employment by older workers as the US population ages.

teams were asked to use these scenarios when their analyses required demographic or economic inputs.

Growing Prosperity

The US economy and population are growing. Barring major wars or other catastrophes, growth is likely to continue through the 21st century. If economic growth is higher, society is likely to be more able to take advantage of the opportunities a changing climate presents, and more able to cope with its negative impacts. Wealthier, industrialized societies derive less of their incomes from strongly climate-related activities than more traditional societies. With more technology and infrastructure, wealthy societies also have more resources to support adaptation, and can more easily endure climate-related losses. Within societies, some will very likely face greater burdens or greater opportunities than others. It is also possible that rapid economic growth can increase vulnerability, by increasing pollution (including greenhouse gas emissions), congestion, demand for land and resources, and stresses on natural ecosystems, and possibly their vulnerability to climate change.

Changing Technology

Much of the recent US economic growth has been fueled by new technology. Although technological change can carry significant social and environmental costs, in aggregate it greatly increased Americans' material well being over the 20th century. For example, in the past decade, new information and communication technologies have transformed many activities, bringing increased productivity and new products and services.

Technology affects society's relationship to climate in many ways. It is very likely that technological change will strongly influence the success of any future efforts to control greenhouse gas emissions, and reduce vulnerability to climate change. For example, it is possible that information technology, combined with new cropping methods and advanced crop varieties, will increase farmers' ability to adapt to climate change or variability. Similarly, advances in medicine, public health, and information technology will likely strengthen our abilities in the early detection, prevention, and treatment of disease.

Technology can also increase society's vulnerability to unanticipated extremes of climate. This can happen because modern society is highly interdependent, relying in critical ways on electric power, transportation, and communications systems, all of which can be disrupted by extreme weather events if systems have not been adequately designed to deal with contingencies.

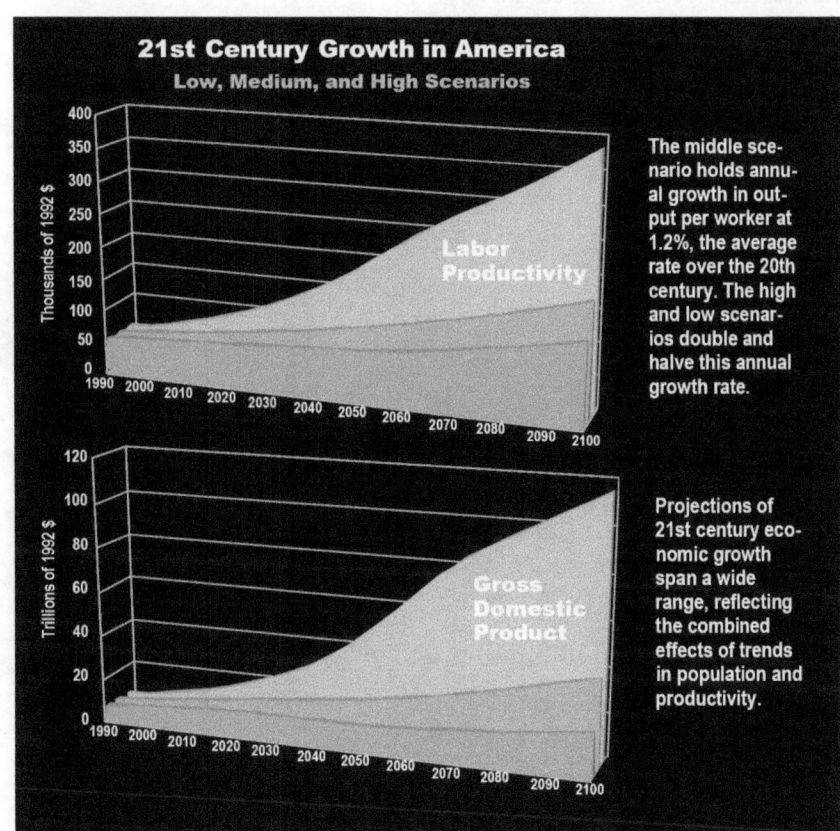

21st Century Growth in America
Low, Medium, and High Scenarios

Labor Productivity

The middle scenario holds annual growth in output per worker at 1.2%, the average rate over the 20th century. The high and low scenarios double and halve this annual growth rate.

Gross Domestic Product

Projections of 21st century economic growth span a wide range, reflecting the combined effects of trends in population and productivity.

OUR CHANGING NATION

A Growing, Aging, and Mobile Populace

Some 53% of the population now live in the 17% of the land area that comprises the coastal zone, and the largest population growth for several decades is projected for coastal areas.

The US population is projected to continue growing through the 21st century, but at a declining rate. The scenarios used in this Assessment project a US population in 2100 that ranges from 353 to 640 million (representing average annual growth rates of 0.31% to 0.86% over the 21st century), with 494 million in the middle scenario. Most of this uncertainty arises from alternative immigration assumptions.

The US population is aging. Over the 20th century, the fraction of Americans over age 65 increased from 1 in 25 to 1 in 8. Older people are physiologically more vulnerable to heat stress. Without adaptive measures, a warmer climate would likely bring an increase in heat-related illness and death, which society's aging would compound. There is also some chance that warming would reduce cold-related mortality, a trend that would also interact with the aging of the population, although the data suggest a weaker effect than for heat. Many older Americans prefer warmer climates, as the migration from northern regions to the Sunbelt demonstrates. Widespread use of one technology, air conditioning, powerfully advanced the growth of these southern regions. At the same time, rapid growth in arid regions has sharply increased these regions' vulnerability to water shortages.

US Population and Growth Trends
Change in county population, 1970-2030

Projected change in county population (percent), 1970 to 2030

- >+250% (highest +3,877%)
- +50% to +250%
- +5% to +50%
- -5% to +5%
- -20% to -5%
- -40% to -20%
- <-40% (lowest -60%)

Each block on the map illustrates one county in the US. The height of each block is proportional to that county's population density in the year 2000, so the volume of the block is proportional to the county's total population. The color of each block shows the county's projected change in population between 1970 and 2030, with shades of orange denoting increases and blue denoting decreases. The patterns of recent population change, with growth concentrated along the coasts, in cities, and in the

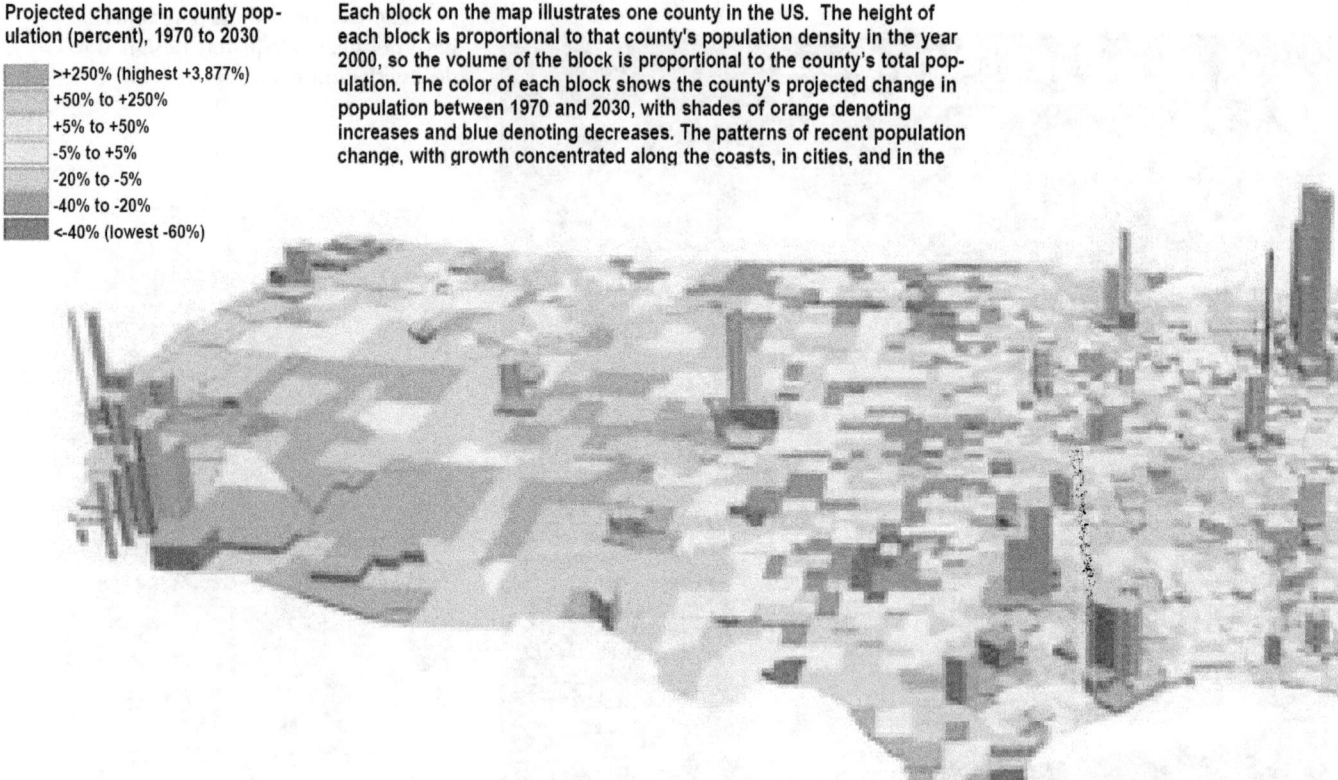

America is becoming more urban. The fraction of Americans living in cities increased from 40% in 1900 to more than 75% today and this increase is projected to continue. Urbanization affects vulnerability to climate and the capacity to adapt in complex ways. City dwellers are less dependent on climate-sensitive activities for their livelihoods, and have more resources and social support

population now live in the 17% of the land area that comprises the coastal zone, and the largest population growth for several decades is projected for coastal areas. Over the next 25 years, population growth of some 18 million is projected in the coastal states of Florida, California, Texas, and Washington. This trend is exacerbating wetland loss and coastal pollution. In addition, locating more and property in low-lying areas increases vulnerability s, storm surges, erosion, and l rise – as several decades of trends, and extreme recent Florida, Georgia, and the s, all confirm.

ng about the Future: with Complexity

A ost of other factors are also ely to affect the ease with ociety can adapt to, or take ge of, climate variability and For particular regions or sec- factors likely to shape cli- lnerability include local zon- nances, housing styles, build- s, popular forms of recre- e age and degree of special- of capital in particular indus- rld market conditions, and ribution of income. To further ate matters, many of these

factors are likely to be influenced by climate variability and change, and to influence each other. Trying to project all such relevant factors, or to model their interactions, would be impossible.

Rather, this first Assessment took a highly simplified approach to projecting socioeconomic factors. When teams needed more detailed socioeconomic projections than the scenarios of population and economic growth provided, they were asked to follow a standard procedure to generate and document the projections they needed. They were asked to select one or two additional factors – such as development patterns, land use, technology, or market conditions – that they judged likely to have the most direct effect on the issue they were examining, and to vary these factors through an uncertainty range they judged plausible. This approach has clear limitations. In fact, teams found the complexity of even this simplified approach challenging, and made limited use of it beyond the basic scenarios. It has, however, allowed some preliminary investigations of the socioeconomic basis of impacts and vulnerability, which can be refined and extended as assessment methods and experience advance.

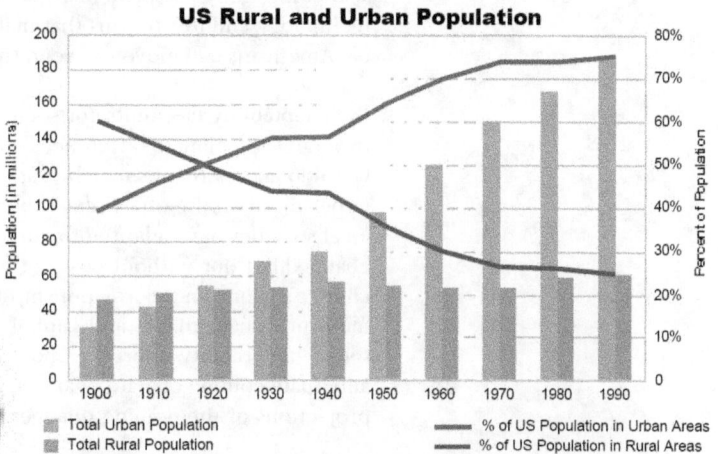

US Rural and Urban Population

Total Urban Population
Total Rural Population
% of US Population in Urban Areas
% of US Population in Rural Areas

Essentially all 20th century US population growth has been in cities, increasing the urban population fraction from 40% in 1900 to more than 75% in 1990. This move to the cities is projected to continue.

OUR CHANGING NATION

Impacts, Adaptation, and Vulnerability

While societies have shown substantial adaptability to climate variability, the challenge of adapting to a climate that is not stable, but evolving at an uncertain rate, has never been tested in an industrialized society.

Climate impacts, vulnerability, and adaptation are distinct but related concepts. Given an assumed state of America's society and economy, the impacts of a specified climate scenario are the differences it yields relative to a continuation of the present climate. These impacts may be beneficial or harmful, with most climate scenarios bringing mixed effects: benefits to some people, places, and sectors, and harms to others. A system is more or less sensitive to climate depending on whether a specified change in climate brings large or small impacts.

People need not merely suffer the climate conditions they face, however, but can change their practices, institutions, or technology to take maximum advantage of the opportunities the climate presents and limit the harms they suffer from its variations. Through such adaptations, people and societies adjust to the average climate conditions and the variability of conditions they have experienced in the recent past. When habits, livelihoods, capital stock, and management practices are finely tuned to current climate conditions, the direct effect of many types of change in these conditions, particularly if the change occurs rapidly, is more likely to be harmful and disruptive than beneficial.

But just as societies adapt to the present climate, they can also adapt to changes in it. Adaptation can be intentional or not, and can be undertaken either in anticipation of projected changes or in reaction to observed changes. Society's capacity to adapt to future climate change is a crucial uncertainty in determining what the actual consequences of climate change will be. Societies and economies are vulnerable to climate change if they face substantial unfavorable impacts, and have limited ability to adapt. Socioeconomic conditions such as wealth, economic structure, settlement patterns, and technology play strong roles in determining vulnerability to specified climate conditions, as the history of US hurricane losses shows.

Human societies and economies have demonstrated great adaptability to wide-ranging environmental and climatic conditions found throughout the world, and to historical variability. Wealthy industrial societies like the US function quite similarly in such divergent climates as those of Fairbanks, Alaska and Orlando, Florida. While individual adaptability also contributes, it is principally social and economic adaptations in infrastructure, capital, technology, and institutions that make life in Orlando and Fairbanks so similar that individual Americans can move between them easily.

But adaptability has limits, for societies as for individuals, and individuals' ability to move through large climate differences tells us little about these limits. Moving between Orlando and Fairbanks may be easy for an individual, but rapidly imposing the climate of either place on the other would be very disruptive. The countless ways that particular local societies have adapted to current conditions and their history of variability can be changed, but not without cost, not all with equal ease, and not overnight. The speed of climate change, and its relationship to the speed at which skills, habits, resource-management practices, policies, and capital stock can change, is consequently a crucial contributor to vulnerability. Moreover, however wisely we may try to adjust long-lived decisions to anticipate coming climate changes, we will inevitability remain limited by our imperfect projections of the coming changes. Effective adaptation may depend as much on our ability to devise responses that are robust to various possible changes, and adjustable as we learn more, as on the quality of our projections at any particular moment. While societies have shown substantial adaptability to climate variability, the challenge of adapting to a climate that is not stable, but evolving at an uncertain rate, has never been tested in an industrialized society.

Consequently, while adaptation measures can help Americans reduce harmful climate impacts and take advantage of associated opportunities, one cannot simply assume that adaptation will make the aggregate impacts of climate change negligible or beneficial. Nor can one assume that all available adaptation measures will necessarily be taken; even for such well-known hazards as fire, flood, and storms, people often fail to take inexpensive and easy risk-reduction measures in their choices of building sites, standards, and materials, sometimes with grave consequences. In this first Assessment, potential climate adaptation options were identified, but their feasibility, costs, effectiveness, or the likely extent of their actual implementation were not assessed. Careful assessment of these will be needed.

Hurricanes and their Impacts in the 20th Century
1900 - 1995

Total Property Losses

Millions of dollars, Constant $1992

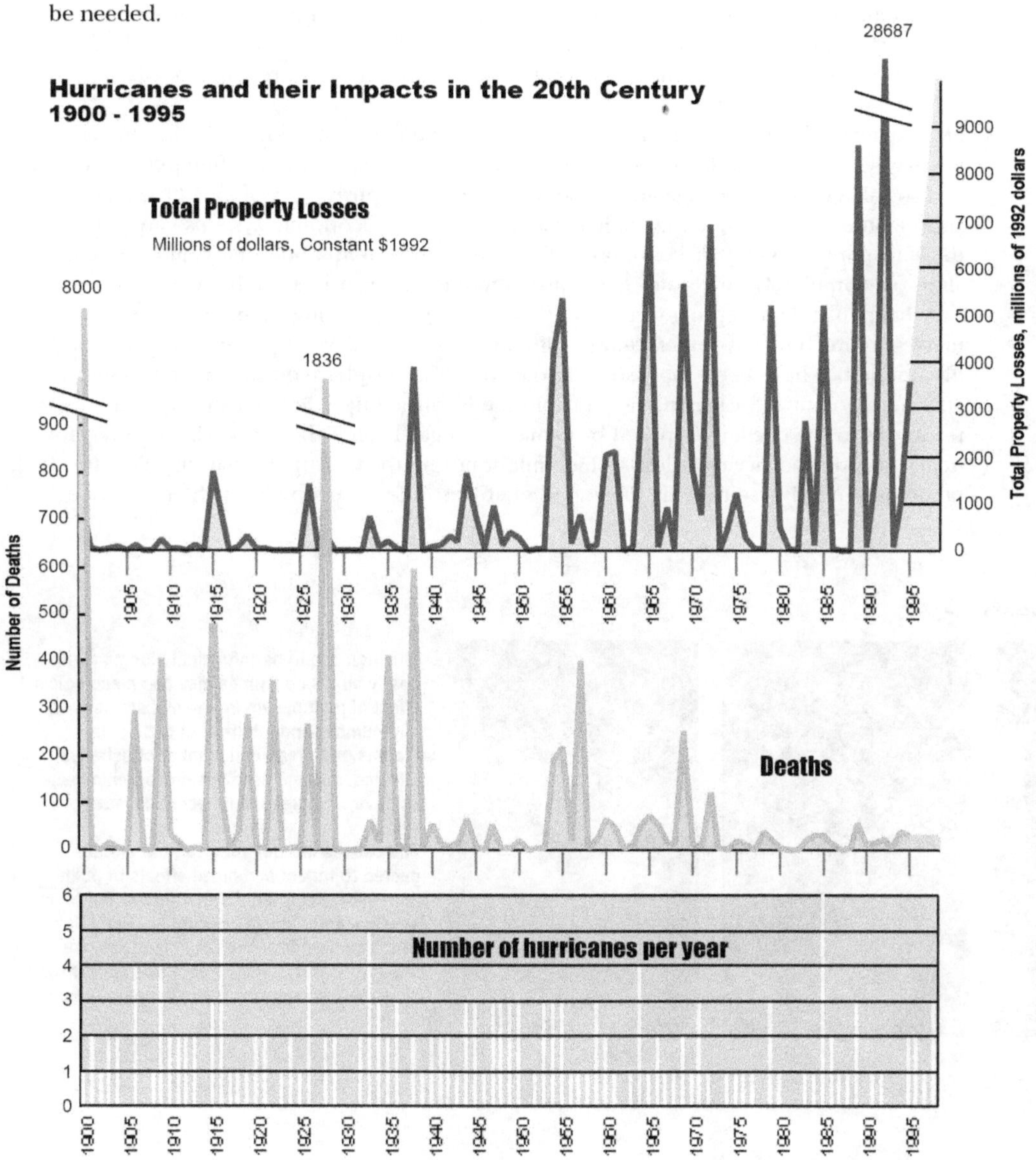

The history of US losses due to hurricanes over the 20th century illustrates the importance of socioeconomic factors in determining vulnerability. Since 1900, economic losses due to hurricanes have increased markedly, while deaths due to hurricanes have decreased markedly – even though there has been no significant trend in the number or intensity of hurricanes. The trends in economic losses and deaths are explained primarily by socioeconomic factors. The decline in deaths reflects the importance of improved forecasts, warning systems, and emergency preparedness measures, an important set of adaptation measures to extreme weather events. The increase in property losses reflects the increasing concentration of valuable property and infrastructure in low-lying coastal regions in the path of hurricanes. Many more people and much more property are now located in harms way, and while we have grown much better at protecting the people from hurricanes, we cannot protect the property.

35

OUR CHANGING NATION

Multiple Stresses, Surprises, and Advancing Knowledge

Past environmental suprises have included the appearance of the 1930s drought, and the 1980s appearance of the Antarctic ozone hole. Potential large-consequence surprises present some of the more worrisome concerns raised by climate change, and pose the greatest challenges for policy and research.

Climate change will occur together with many other economic, technological, and environmental trends, which may stress the same ecological and social systems and interact with climatic stresses. Human society has imposed various stresses on the environment, at diverse scales, for centuries. Over the 21st century some non-climatic stresses will likely increase (such as loss of habitat) while others decrease (such as acidifying pollution); climate change is likely to compound some non-climatic stresses and mitigate others. Systems that are already bearing multiple other stresses are likely to be more vulnerable to climatic stress. This applies to communities and managed ecosystems, such as marginal agriculture or resource-based communities suffering job loss and out-migration. It also applies to natural ecosystems, whose capacity for adaptation is, in general, likely to be much more limited than that of human communities. Although the central importance of considering interactions between multiple stresses is clear, present tools and methods for doing this are limited; this limitation points to an important set of research needs.

Many climate changes and their impacts will likely be extensions of trends that are already underway, and so are at least partly predictable, but some are not. We often expect natural and social systems to change and respond continuously: push the system a little, and it shifts a little. But complex climatic, ecological, and socioeconomic systems can sometimes respond in highly discontinuous ways: push the system a little more, and it might shift to a completely new state. Such discontinuities or surprises can be seen clearly after they happen, and attempting to explain them often generates important advances in our understanding, but they are extremely difficult to predict. Several possible surprises and discontinuities have been suggested for the Earth's atmosphere, oceans, and ecosystems. Still more potential for surprise arises from the intrinsic unpredictability of human responses to the challenges posed by climate change. Even if the probability of any particular surprise occurring is low (which might not be the case), potential surprises are so numerous and diverse that the likelihood of at least one occurring is much greater. We

Harmful Algal Blooms

California

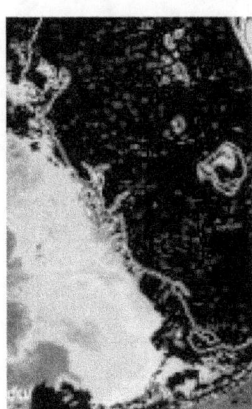

Florida

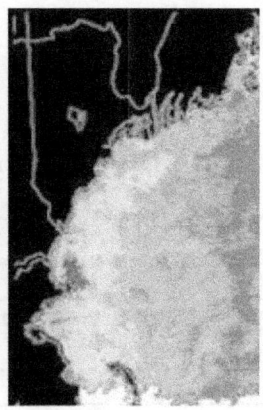

New England

The increase in harmful algal blooms along nearly all US coastlines may be an ecological effect of multiple environmental stresses. The number and intensity of toxic algal blooms, the areas and number of fisheries affected, and the associated economic losses have all increased in recent decades.

The causes are not yet clear, but are suspected to reflect combined effects of pollution, excess nutrient input, transport of toxic species, and climate conditions. Red tides, which can render shellfish poisonous, are one type of toxic algal bloom that appear to increase with warmer ocean temperatures.

have been surprised by environmental and socioeconomic changes many times. Examples of past environmental surprises include the appearance of the 1930s drought, and the 1980s appearance of the Antarctic ozone hole. Potential large-consequence surprises present some of the more worrisome concerns raised by climate change, and pose the greatest challenges for policy and research.

Surprises are inherently unpredictable. But two broad approaches can help us prepare to live with a changing and uncertain climate, even considering the possibility of surprise. First, some of our assessment effort can be devoted to identifying and characterizing potential large-impact events, even if we presently judge their probability to be very small. Second, society can maintain a diverse and advancing portfolio of scientific and technical knowledge, and policies that encourage the creation and use of new knowledge and technology. This would provide a powerful foundation for adapting to whatever climate changes might come.

Continually advancing knowledge and technology, and the social, economic, and policy conditions that support them, provide a powerful foundation for adapting to whatever climate changes might come.

Climate Change and other Environmental Stresses

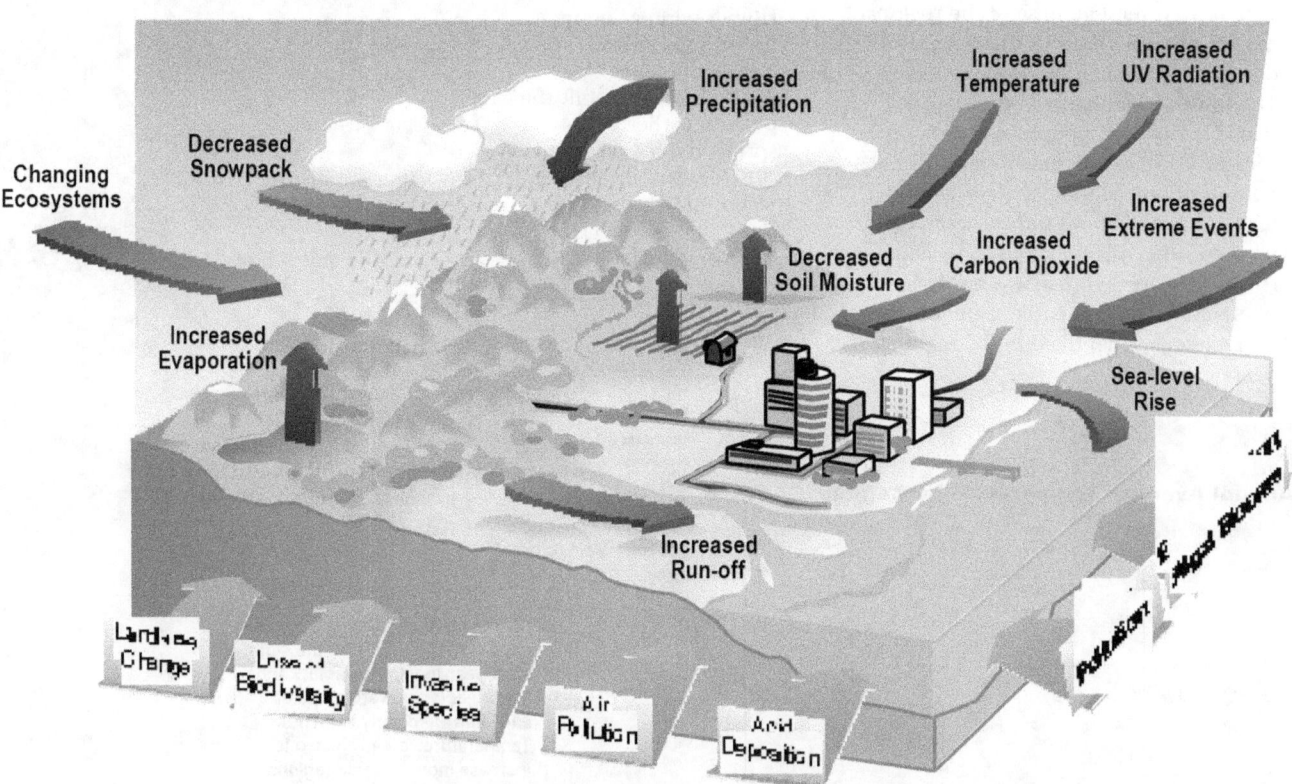

Although climate change has multiple linked impacts, climate change is itself just one of many forms of change underway in the global environment, connected in complex and uncertain ways. The figure illustrates several of the effects of climate change and several other forms of environmental change underway. Developing tools and methods to assess the impacts of climate change in the context of other environmental changes will be a high priority for future assessments.

REGIONAL OVERVIEW

There are both important commonalities and important differences in the climate-related issues and consequences faced around the country. For example, water is a key issue in virtually all regions, but the specific changes and impacts in the West, in the Great Lakes, and the Southeast will differ. Regional texture is thus critical in thinking through how to best respond to the changing climate we will face in the coming decades and century.

Twenty regional workshops involving a wide range of researchers and stakeholders helped identify key issues facing each region and began identifying potential adaptation strategies. This report groups the findings of these efforts into larger regions to offer a glimpse of the regional mosaic of consequences that are possible due to climate change and variability. The impacts highlighted here suggest that it is vital that people everywhere start to learn about climate impacts and consider them in their short- and long-term decisions about infrastructure, land use, and other planning. In many cases, research is needed to assess the feasibility, effectiveness, and costs of the adaptation strategies identified in the regional overviews.

Alaska

Sharp winter and springtime temperature increases are very likely to cause continued thawing of permafrost, further disrupting forest ecosystems, roads, and buildings.

Northwest

Increasing stream temperatures are very likely to further stress migrating fish, complicating restoration efforts.

Mountain West

Higher winter temperatures are very likely to reduce snowpack and peak runoff and shift the peak to earlier in the spring, reducing summer runoff and complicating water management for flood control, fish runs, cities, and irrigation.

Southwest

With an increase in precipitation, the desert ecosystems native to this region are likely to decline while grasslands and shrublands expand.

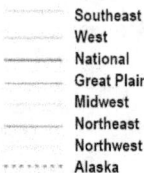

----------- Southeast
----------- West
——————— National
----------- Great Plains
----------- Midwest
——————— Northeast
          Northwest
·········· Alaska

Annual Average Temperature by Region

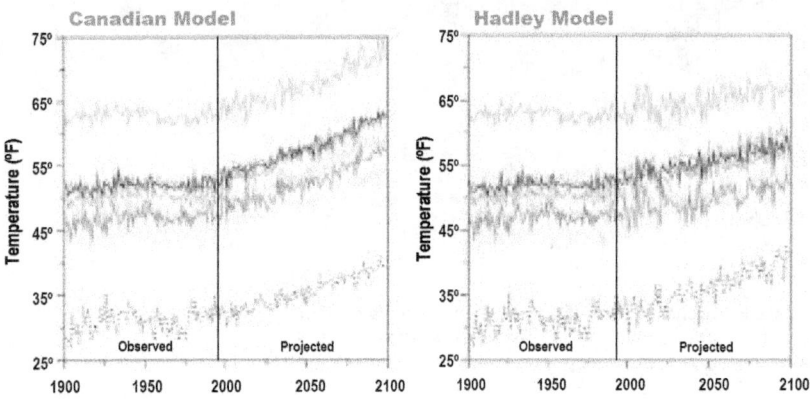

Average temperature is shown for each region in °F. Both the significant year-to-year variability and the projected upward slope of each line are clearly evident. Temperatures are projected to increase more in some regions than in others, with generally larger increases projected by the Canadian model scenario than by the Hadley model scenario.

Midwest/Great Plains

Higher CO_2 concentrations are likely to offset the effects of rising temperatures on forests and agriculture for several decades, increasing productivity.

Great Lakes

Lake levels are likely to decline, leading to reduced water supply and more costly transportation. Shoreline damage due to high water levels is likely to decrease.

Northern and Mountain Regions

It is very probable that warm weather recreational opportunities like hiking will expand while cold weather activities like skiing contract.

Northeast, Southeast, and Midwest

Rising temperatures are very likely to increase the heat index dramatically in summer, with impacts to health and comfort. Warmer winters are likely to reduce cold-related stresses.

Appalachians

Warmer and moister air will very likely lead to more intense rainfall events, increasing the potential for flash floods.

Southeast

Under warmer wetter scenarios, the range of southern tree species is likely to expand. Under hotter and drier scenarios, it is likely that far southeastern forests will be displaced by grasslands and savannas.

Southeast Atlantic Coast

It is very probable that rising sea levels and storm surge will threaten natural ecosystems and human coastal development and reduce buffering capacity against storm impacts.

Great Plains

Prairie potholes, which provide important habitat for ducks and other migratory waterfowl, are likely to dry up in a warmer climate.

Southeast Gulf Coast

Inundation of coastal wetlands will very likely increase, threatening fertile areas for marine life, and migrating birds and waterfowl.

Islands

More intense El Niño and La Niña events are possible and are likely to create extreme fluctuations in water resources for island citizens and the tourists who sustain local economies.

Great Plains

Midwest

Northeast

Southeast

Islands

Hawaii

Pacific   Carribean

NORTHEAST

The Northeast is characterized by diverse waterways, extensive shorelines, and a varied landscape. The contrasts, from mountain vistas and extensive forests to one of the most densely populated corridors in the US, are noteworthy. The Northeast includes the largest financial market in the world (New York City), the nation's most productive non-irrigated agricultural county (Lancaster, PA), and the largest estuarine region (the Chesapeake Bay) in the US. The Northeast is dominated by managed vegetation, with much of the landscape covered by a mosaic of farmland and forest. The varied physical setting of the Northeast is matched by its highly diversified economy. The majority of the population is concentrated in the coastal plain and piedmont regions, and in major urban areas. Economic activities in the region include agriculture, resource extraction (forestry, fisheries, and mining), major service industries highly dependent on communication and travel, recreation and tourism, and manufacturing and transportation of industrial goods and materials.

Observed Climate Trends

Historically, the Northeast has experienced significant variability and extreme events related to weather and climate. Floods, droughts, heat waves, and severe storms are characteristic. For example, seven major tropical storms have crossed the mid-Atlantic region since 1986 and six years of the last 20 have been characterized by significant drought. In addition, the major cities of the Northeast have experienced episodes of increased illness and deaths during heat waves. Temperature increases of as much as 4°F (2°C) over the last 100 years have occurred along the coastal margins from the Chesapeake Bay through Maine. Precipitation has generally increased, with trends greater than 20% over the last 100 years occurring in much of the region. Precipitation extremes appear to be increasing while the amount of land area experiencing drought appears to be decreasing. For the region as a whole, the period between the first and last dates with snow on the ground has decreased by 7 days over the last 50 years.

KEY ISSUES

- Increase in Weather Extremes

- Stresses on Estuaries, Bays, and Wetlands

- Multiple Stresses on Urban Areas

- Recreation Shifts

- Human Health

- Species Changes

The Vulnerability of Urban Transportation Systems

Historical events often illustrate vulnerabilities. The December 11-12, 1992 nor'easter produced some of the worst flooding and strongest winds on record for the area. It resulted in a near shutdown of the New York metropolitan transportation system and evacuation of many seaside communities in New Jersey and Long Island. This storm provided a "wake-up" call, indicating the vulnerability of the transportation system to major nor'easters and hurricanes. Critical transportation systems are only 7 to 20 feet above current sea level. Had flood levels been only 1 to 2 feet above the actual 8.5 foot high water, massive inundation of rail and subway tunnels could have resulted, with possible loss of life. There is a possibility that sea-level rise due to climate change will add 1 to 3 feet to all surge heights by 2100, so even a weaker storm would produce damage comparable to the 1992 storm. The construction of dikes and pumping stations and the institution of effective warning systems are possible

Scenarios of Future Climate

The Northeast has among the lowest rates of projected future warming compared to other regions of the US. Winter minimum temperatures show the greatest change,with projected increases ranging from 4-5°F (2-3°C) to as much as 9°F (5°C) by 2100,with the largest increases in coastal regions. Maximum temperatures are likely to increase much less than minimums, again, with the largest changes in winter. Model scenarios offer a range of potential future changes for precipitation,from roughly 25% increases by 2100,to little change or small regional decreases. The variability in precipitation in the coastal areas of the Northeast is projected to increase. Models provide contrasting scenarios for changes in the frequency and intensity of winter storms.

The view of Mount Washington in New Hampshire changes dramatically between a clear day (top photo) and a day when temperatures exceeding 90˚F exacerbate air quality problems across the region.

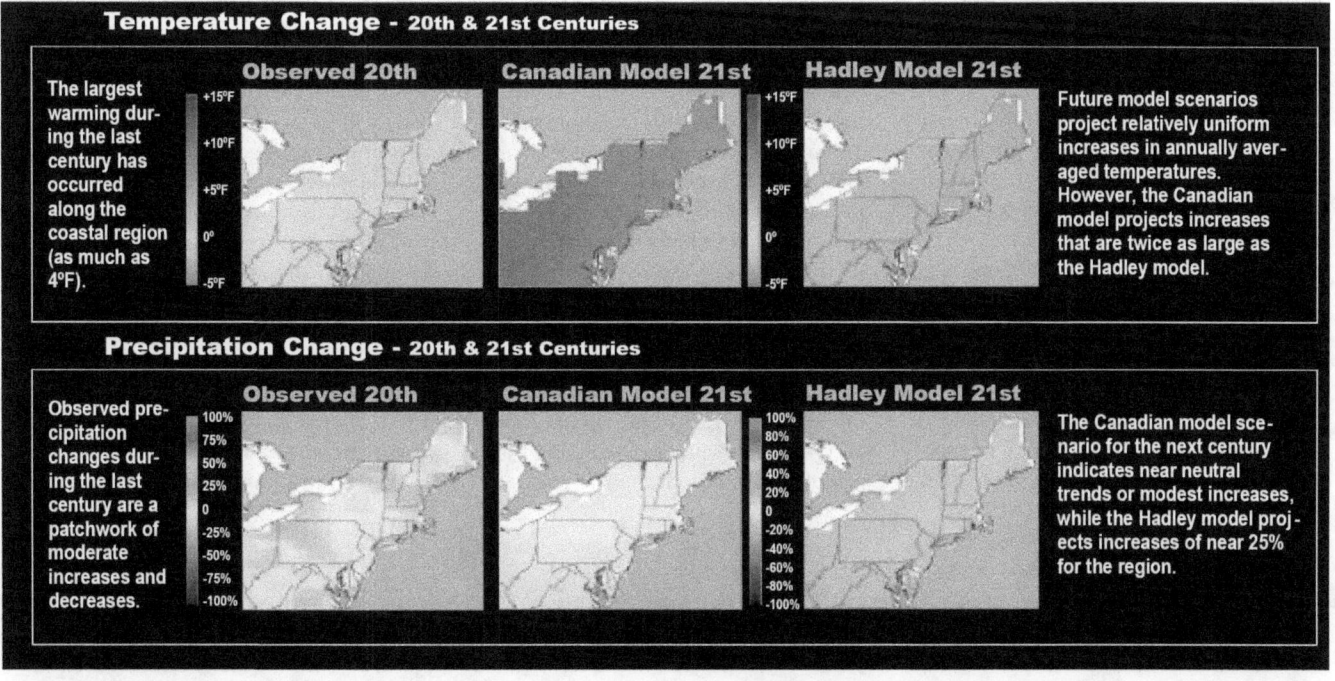

Temperature Change - 20th & 21st Centuries

The largest warming during the last century has occurred along the coastal region (as much as 4°F).

Observed 20th	Canadian Model 21st	Hadley Model 21st
+15°F +10°F +5°F 0° -5°F		+15°F +10°F +5°F 0° -5°F

Future model scenarios project relatively uniform increases in annually averaged temperatures. However, the Canadian model projects increases that are twice as large as the Hadley model.

Precipitation Change - 20th & 21st Centuries

Observed precipitation changes during the last century are a patchwork of moderate increases and decreases.

Observed 20th	Canadian Model 21st	Hadley Model 21st
100% 75% 50% 25% 0 -25% -50% -75% -100%		100% 80% 60% 40% 20% 0 -20% -40% -60% -80% -100%

The Canadian model scenario for the next century indicates near neutral trends or modest increases, while the Hadley model projects increases of near 25% for the region.

adaptation strategies. While hurricanes are much less frequent than nor'easters in the Metro East region, they can be even more destructive because the geometry of the coast amplifies surge levels toward the New York City harbor. For a worst-case scenario category 3 hurricane, surge levels could rise 25 feet above mean sea level at JFK Airport and 21 feet at the Lincoln Tunnel entrance.

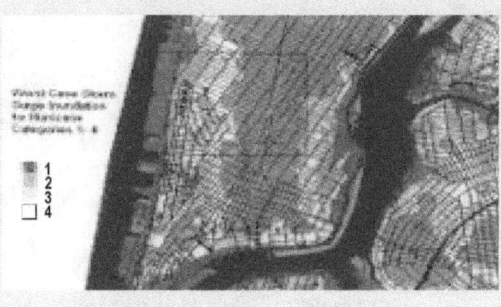

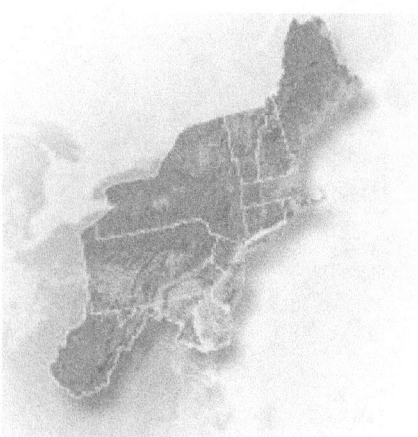

NORTHEAST KEY ISSUES

Increase in Weather Extremes

The Northeast is prone to natural weather extremes and disasters including ice storms, severe flooding, nor'easters, hurricanes, and severe or persistent drought. The ice storm of January 1998, with extensive tree damage and an extended period of power failure, the severe flooding associated with tropical depression Floyd in 1999, and six significant droughts over the last 20 years amply demonstrate the importance of weather extremes to the region.

Climate change is likely to decrease the number of some types of weather extremes, while increasing others. The warming projected by climate models over the next several decades suggests possible increases in rain events over frozen ground or rapid snow melting events that can increase flooding. Over the coming century, winter snowfalls and periods of extreme cold will likely decrease. In contrast, heavy precipitation events have been increasing and warming is likely to continue this trend. Potential changes in the intensity and frequency of hurricanes are a major concern.

Adaptations: Possible strategies include relocating or elevating structures that are at risk from severe weather and flooding, though this may not be practical in many cases. Enhanced design of critical infrastructure (such as power supply) may improve the likelihood of continuous operation during extreme weather events. The complex institutional framework of community, municipal, county, regional, and statewide formal and informal governing bodies and infrastructure of the Northeast have the potential to limit the region's ability to respond to extreme events. Although there are signs of innovative management strategies, the ability of the Northeast to adapt to extreme situations will depend upon the ability of institutions to identify and prioritize vulnerable facilities and populations. The large differences in economic status and the aging of the population in the Northeast are also likely to be associated with differential impacts based on the ability to respond to climate change. Where impacts are significant, climate change is likely to have greater impact on lower-income residents, the elderly, children, and the ill (such as those with chronic respiratory ailments).

Stresses on Estuaries, Bays, and Wetlands

The estuaries, bays, and wetlands of the Northeast coastal zone are highly valued as unique ecosystems, major recreational sites, migratory waterfowl habitats, and fishery sources. The largest US estuary is the Chesapeake Bay. The bay is heavily stressed by air and water pollution from industry, agriculture, and cities. Farm and urban runoff carries particles, as well as fertilizer and other excess nutrients into the Bay. These pollutants initiate processes that reduce oxygen levels in the water. Climate change is likely to exacerbate these stresses by increasing water temperature. Changes in precipitation

The photo shows a car which has been buried in heavy snow. It is possible that over the next few decades, the effects of warming will be counter-intuitive. If Lake Erie and Lake Ontario have shorter seasons of ice cover, it is possible that lake effect snows in cities like Cleveland and Buffalo will increase during mid-winter. Later in the 21st century, snowfall will likely decrease with the greater warming.

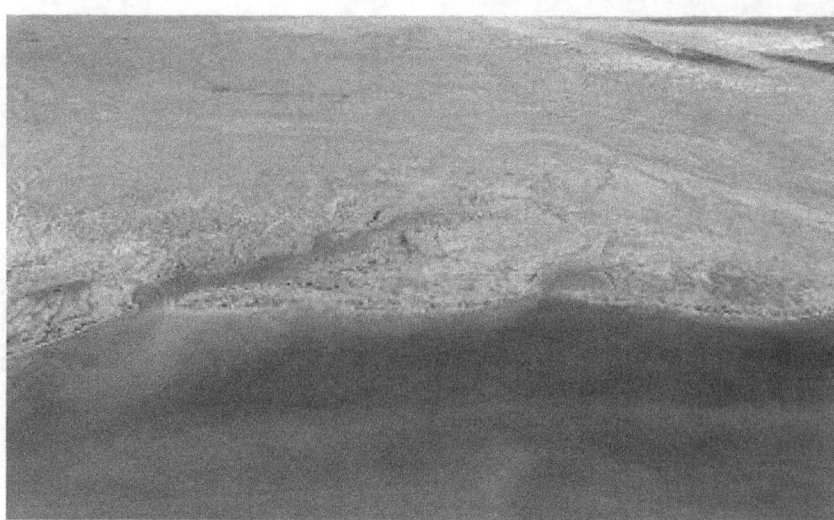

The estuaries of the mid-Atlantic region include the Chesapeake and Delaware Bays. These estuaries are geographically fixed, and so, unlike plant and animal species, they cannot move in response to climate change.

and runoff, much more uncertain elements of the climate scenarios, also affect coastal salinity. Both temperature and salinity have significant effects on fish populations, and human and ecosystem health. Sea-level rise is very likely to substantially increase wetland and marsh loss. Climate change impacts are therefore very likely to compound the many other stresses on the bays and estuaries of the Northeast.

Adaptations: Strategies include measures to reduce the flow of excess nutrients into the bay from industrial, urban, and rural non-point sources, reduce atmospheric deposition, and better enforce existing wetland policies. The overall reduction of stresses from other sources could help to enhance the resiliency of coastal ecosystems to additional and sometimes uncertain stresses from climate variability and change. In addition, acquisition of lands contiguous to coastal wetlands could allow for their inland migration as sea level rises.

Percent Salinity Change in the Chesapeake Bay

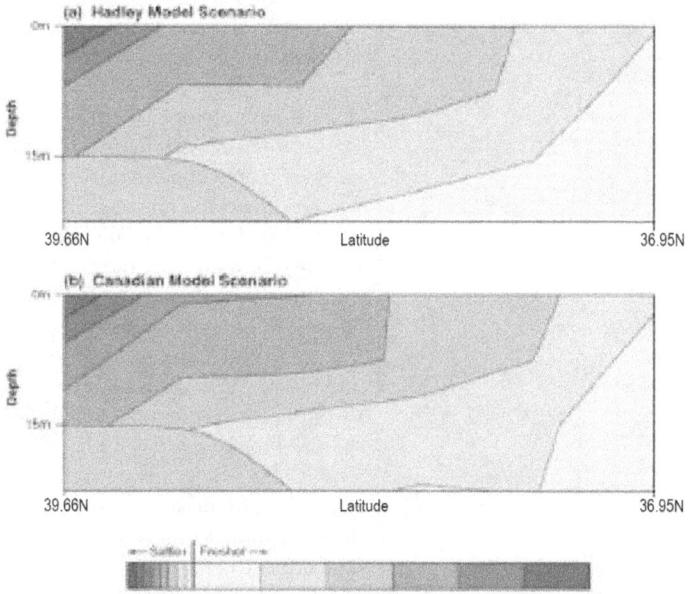

The Hadley (top) and Canadian (bottom) climate scenarios each yield a very different salinity for the Chesapeake Bay based on water balance calculations for the region extending from the upper reaches of the Bay (39.66N latitude) to the Lower Chesapeake near its opening to the Atlantic. Salinity has a significant impact on populations of fish and other organisms.

Winter Minimum Temperature

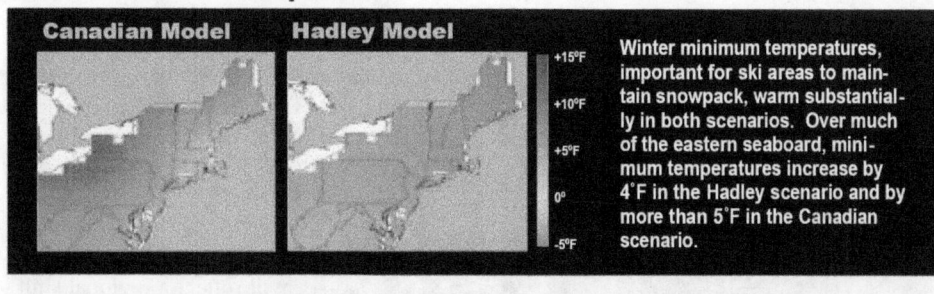

Winter minimum temperatures, important for ski areas to maintain snowpack, warm substantially in both scenarios. Over much of the eastern seaboard, minimum temperatures increase by 4°F in the Hadley scenario and by more than 5°F in the Canadian scenario.

July Heat Index Change - 21st Century

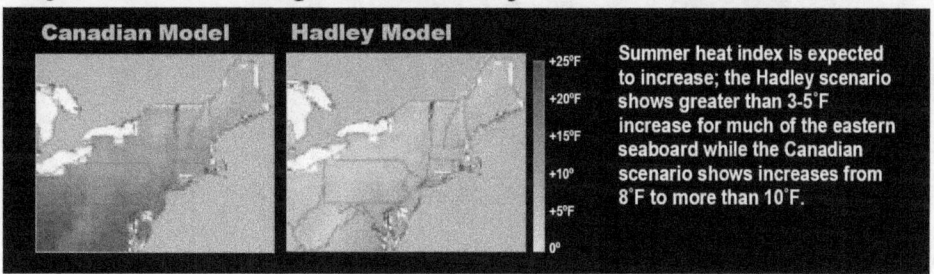

Summer heat index is expected to increase; the Hadley scenario shows greater than 3-5°F increase for much of the eastern seaboard while the Canadian scenario shows increases from 8°F to more than 10°F.

Winter minimum temperatures show the greatest change, with projected increases ranging from 4-5°F (2-3°C) to as much as 9°F (5°C) by 2100.

NORTHEAST KEY ISSUES

Multiple Stresses on Urban Areas

Milder winters contribute to a higher survival rate for deer and mice, both of which are factors in the population of the deer tick, the primary Lyme disease vector.

Recent examples of outbreaks of West Nile Virus and equine encephalitis in northeastern urban areas have substantially raised concerns about vector-borne diseases.

Climate change will very likely intersect with many existing stresses (some climate-related and some not) on the major urban areas of the Northeast, with implications for the overall quality of life. The infrastructure of many major Northeast cities (such as water supply, communication, energy delivery, and waste disposal) is characterized by aging, insufficient capacity, and deferred maintenance. Other existing stresses include crime, chronic air-quality problems, and inadequate power supply to meet peak energy demands. Decreased snowfalls and more moderate winter temperatures are likely to result in decreased winter stresses. However, climate change has greater potential to add to existing stresses. Major potential consequences of climate change include the impacts of rising sea level and elevated storm surges on transportation systems, increased heat-related illness and death associated with temperature extremes, increased ground-level ozone pollution associated with warming, and the impact of precipitation and evaporation changes on relatively inflexible water supply systems.

Adaptations: Strategies include changing water supply management; replacing aging infrastructure with more climate-resilient systems; strengthening water quality and air quality controls to minimize the compounding of climate impacts; and using early warning systems and measures such as changing roofing colors and adding shade trees to limit urban heat that can contribute to heat-related stresses and deaths.

Recreation Shifts

Increased warmth and changes in the seasonal characteristics of precipitation are likely to have substantial impacts on recreation in the Northeast. Typical summer recreational activities involving beaches or freshwater reservoirs are likely to have extended seasons, with the region's diverse waterways becoming havens for escape from increasing summer heat. Possible negative impacts include limiting the ability of ski areas to maintain snow pack, muting of fall foliage colors, increases in insect populations, and worsening ground-level ozone pollution problems, even in the mountains of New England. Higher sea level coupled with even moderate storms will probably result in loss of beachfront property.

Adaptations: Strategies will reflect a regional shift in recreational activity as people make trade-offs in terms of the type, location, and season of their activities.

Warmer winters are likely to limit the viability of snow skiing in the Northeast.

Human Health

Populations of infectious disease vectors are often influenced by climate. Altered mosquito populations and Lyme disease vectors are possible changes in response to higher temperatures (particularly the milder winters projected by virtually all climate models) and changes in moisture. Milder winters contribute to a higher survival rate for deer and mice, both of which are factors in the population of the deer tick, the primary Lyme disease vector. However, the complexity of the relationships makes changes in the distribution and frequency of the disease under altered climate difficult to predict. While warmer and wetter conditions may alter insect vector survival, research is lacking on how these changes may influence disease occurrence. The recent examples of outbreaks of West Nile Virus and equine encephalitis in northeastern urban areas have substantially raised concerns about vector-borne diseases and illustrate that improved monitoring and better understanding of these diseases are relevant for the region. Increased rainfall and flooding, if severe, creates conditions for possible public and private water source contamination (such as with

Cryptosporidium). However, in large measure, US public health infrastructure and response capabilities, if vigorously sustained, are likely to limit many potential impacts.

Species Changes

Changes in species composition are often associated with changes in temperature and precipitation. Key concerns involve the potential for changes in predator-prey relationships, changes in pest types and populations, invasive species, and in key species that are truly characteristic of a region or are of economic significance. For example, lobster populations are associated with cooler waters and warming is thus likely to promote northward migration of the lobster population – a key issue for New England. Coastal population pressures combined with sea-level rise are very likely to reduce habitat for migratory birds along the Atlantic Flyway. Warming is also likely to substantially limit trout populations – a key issue for Pennsylvania. Changes in species mix and introduction of climate-driven invasive species are likely to also induce unanticipated feedbacks on ecosystems.

The likely migration of sugar maple trees northward into Canada as climate warms would sharply reduce maple syrup production, a cultural tradition in the Northeast.

The diverse waterways and extensive forests of the Northeast are likely to have more warm weather recreational use.

Lobster populations are associated with cooler waters and warming is thus likely to promote northward migration of the lobster population – a key issue for New England.

Dominant Forest Types

- White-Red-Jack Pine
- Spruce-Fir
- Longleaf-Slash Pine
- Loblolly-Shortleaf Pine
- Oak-Pine
- Oak-Hickory
- Oak-Gum-Cypress
- Elm-Ash-Cottonwood
- Maple-Beech-Birch
- Aspen-Birch
- No Data

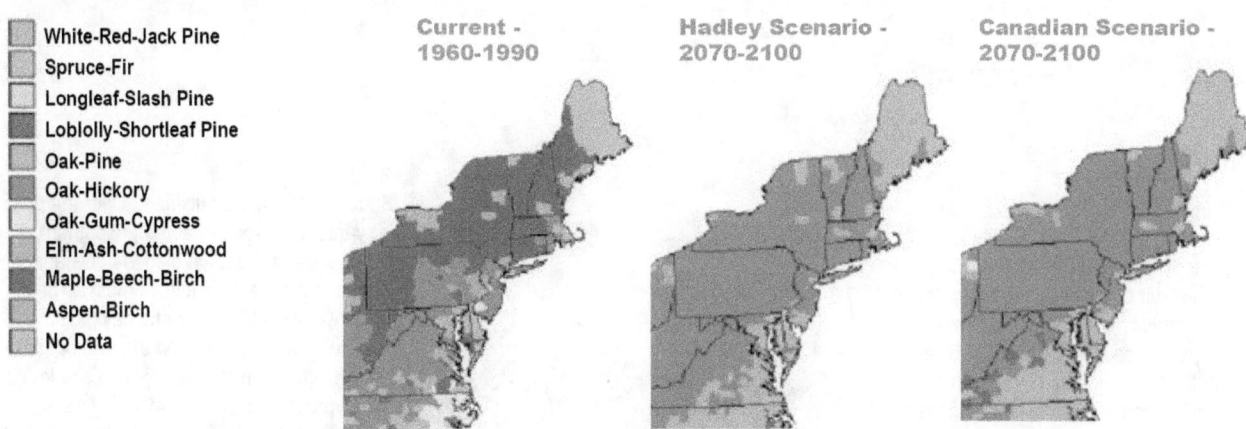

Current - 1960-1990

Hadley Scenario - 2070-2100

Canadian Scenario - 2070-2100

The maps above show current and projected forest types for the Northeast, based on the DISTRIBmodel (see Forest sector). Note that Maple-Beech-Birch, currently a dominant forest type in the region, is completely displaced by other forest types in both the Hadley and Canadian climate scenarios.

SOUTHEAST

The Southeast "sunbelt" is a rapidly growing region with population increasing by more than 30% between 1970 and 1990. Much of this growth occurred in coastal counties,which are projected to grow another 40% between 2000 and 2025. The number of farms in the region decreased 80% between 1930 and 1997,but the Southeast still produces roughly one quarter of US agricultural crops. The Southeast has become America's "woodbasket," producing about half of America's timber supplies. The region also produces a large portion of the nation's fish,poultry,tobacco,oil,coal,and natural gas. Prior to European settlement,the landscape was primarily forests, grasslands,and wetlands,but most of the native forests were converted to managed forests and agricultural lands by 1920. Roughly half of the remaining wetlands in the lower 48 states are located in the Southeast,and more than three-quarters of the Nation's annual wetland losses over the past 50 years occurred in this region. Although much of the landscape has been altered,a wide range of ecosystem types exists and overall species diversity is high.

Observed Climate Trends

Temperature trends in the Southeast vary between decades,with a warm period during the 1920s-1940s followed by a cooling trend through the 1960s. Since the 1970s,temperatures have been increasing,with the 1990's temperatures as warm as the peaks in the 1920s and 30s. Annual rainfall trends show very strong increases of 20-30% or more over the past 100 years across Mississippi,Arkansas,South Carolina,Tennessee,Alabama,and parts of Louisiana, with mixed changes across most of the remaining area. There has been a strong tendency for more wet spells in the Gulf Coast states,and a moderate tendency in most other areas. The percentage of the Southeast landscape experiencing severe wetness increased approximately 10% between 1910 and 1997. There are strong El Niño and La Niña effects in the Southeast that can result in dramatic seasonal and year-to-year variations in temperature and precipitation. El Niño events also tend to create atmospheric conditions that inhibit Atlantic tropical storm development, resulting in fewer hurricanes. La Niña events have the opposite effect, resulting in more hurricanes.

Ghost Forests

Vast stands of coastal forest are dying along the Gulf of Mexico shoreline. Sea-level rise resulting in saltwater intrusion is the suspected cause, and the sun-bleached remnants of dead stems have given rise to the common term "ghost forest" in parts of

South Florida and Louisiana. Over the past 30 years, hundreds of acres of southern baldcypress trees have died in Louisiana coastal parishes, with losses most acute in areas where subsidence and navigation channels have accelerated the rate of saltwater encroachment due to rising sea level. Baldcypress and live oak mortality have occurred as far as 30 miles inland. In

Scenarios of Future Climate

C limate model projections exhibit a wide range of plausible scenarios for both temperature and precipitation over the next century. Both of the principal climate models used in the National Assessment project warming in the Southeast by the 2090s, but at different rates. The Canadian model scenario shows the Southeast experiencing a high degree of warming, which translates into lower soil moisture as higher temperatures increase evaporation. The Hadley model scenario simulates less warming and a significant increase in precipitation (about 20%). Some climate models suggest that rainfall associated with El Niño and the intensity of droughts during La Niña phases will be intensified as atmospheric CO_2 increases.

Louisiana's Coastal Land loss
Between 1956 and 1990 (Shown in Red)

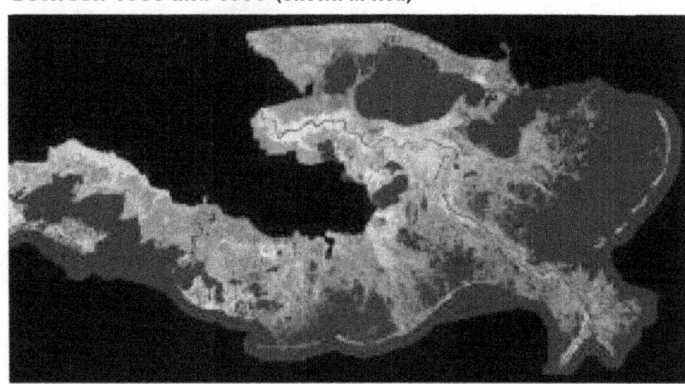

Rising sea level is one of several factors that have caused the loss of about one million acres of Louisiana wetland since 1900. Natural and human-induced processes contributing to these losses include subsidence due to groundwater withdrawal and natural sediment compaction, wetland drainage, and levee construction. The white line designates the coastal zone and red designates land that has been converted to open water.

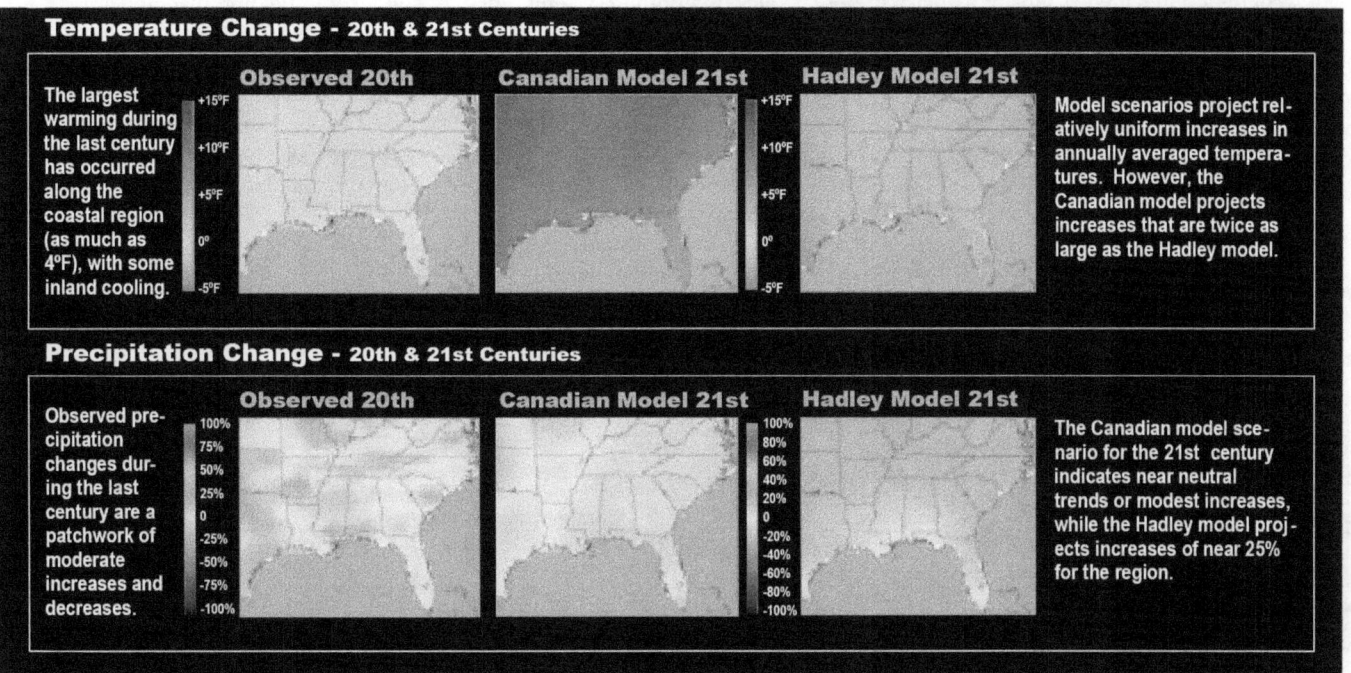

Temperature Change - 20th & 21st Centuries

Observed 20th | **Canadian Model 21st** | **Hadley Model 21st**

The largest warming during the last century has occurred along the coastal region (as much as 4°F), with some inland cooling.

+15°F / +10°F / +5°F / 0° / -5°F

Model scenarios project relatively uniform increases in annually averaged temperatures. However, the Canadian model projects increases that are twice as large as the Hadley model.

Precipitation Change - 20th & 21st Centuries

Observed 20th | **Canadian Model 21st** | **Hadley Model 21st**

Observed precipitation changes during the last century are a patchwork of moderate increases and decreases.

100% / 75% / 50% / 25% / 0 / -25% / -50% / -75% / -100%

100% / 80% / 60% / 40% / 20% / 0 / -20% / -40% / -60% / -80% / -100%

The Canadian model scenario for the 21st century indicates near neutral trends or modest increases, while the Hadley model projects increases of near 25% for the region.

Florida, chronic saltwater contamination of forest soils occurs nearer the shoreline.

Since 1991 landowners and public land managers in Florida have observed massive die-offs of sabal palm along a 40-mile stretch of coast between Cedar Key and Homosassa Springs. Ed Barnard, a forest pathologist with Florida's Forestry Division,

compares what he has seen with the aftermath of Hurricane Hugo in South Carolina, and he attributes the Florida problem to saltwater.

Analyses also attribute the forest decline to salt water intrusion associated with sea-level rise. Since 1852, when the first topographic charts of this region were prepared,

high tidal flood elevations have increased approximately 12 inches. Coastal forest losses will be even more severe if sea-level rise accelerates as is expected as a result of global warming.

SOUTHEAST KEY ISSUES

Weather-related Stresses on Human Populations

The Southeast is prone to frequent natural weather disasters that affect human life and property. Over half of the nation's costliest weather-related disasters of the past 20 years have occurred in the Southeast, costing the region over $85 billion in damages, mostly associ-

Flooded community along Bayou Lafourche in South Louisiana after landfall of Hurricane Juan in 1985.

ated with floods and hurricanes. Across the region, intense precipitation has increased over the past 100 years, and this trend is projected to continue.

The southern heat wave and drought of 1998 resulted in damages in excess of $6 billion and at least 200 deaths. Human health concerns arise from the projected increases in maximum temperatures and heat index in the region. These concerns are particularly great for lower income households that lack sufficient resources to improve insulation and install and operate air conditioning systems. Air quality degradation in urban areas is also a concern associated with elevated air temperatures and increased emissions from power generation, which can increase ground-level ozone. Increased flooding in low-lying coastal counties from the Carolinas to Texas is also likely to adversely impact human health;

floods are the leading cause of death from natural disasters in the region and nationwide.

Adaptations: Traditional approaches such as flood levees, elevated structures, and building codes are no longer adequate by themselves, particularly in the coastal zone, as sea-level rise alone continues to increase the propensity for storm-surge flooding in virtually all southeastern coastal areas. Improvements in risk assessment, coastal and floodplain management, linking insurance to policies for mitigating flood damage, and local mitigation planning are strategies that are likely to decrease potential costs. Changes in climate and sea-level rise should be an integral consideration as coastal communities develop strategies for hazard preparedness and mitigation.

July Heat Index Change - 21st Century

Canadian Model   **Hadley Model**

+25°F
+20°F
+15°F
+10°
+5°F
0°

The changes in the simulated heat index for the Southeast are the most dramatic in the nation with the Hadley model suggesting increases of 8 to 15°F for the southern-most states, while the Canadian model projects increases above 20°F for much of the region.

Summer Soil Moisture Change - 20th & 21st Centuries

Observed 20th   **Canadian Model 21st**   **Hadley Model 21st**

100%
75%
50%
25%
0
-25%
-50%
-75%
-100%

The observed record of soil moisture illustrates mixed changes across the region.

The Hadley model projects soil moisture will increase substantially in most of the region. In contrast, the Canadian model, with larger projected warming and little change in rainfall, suggests large decreases in soil moisture.

Agricultural Crop Yields and Economic Impacts

Crop yield and economic impact estimates vary by climate scenario, area, and crop. The Hadley scenario simulates decreases in the yield of most dryland (non-irrigated) crops in the Gulf Coast area but increases elsewhere in the region through both the 2030s and 2090s. Average yields of irrigated soybean, wheat, and rice increase under the Hadley scenario by 10% in 2030 and by more than 20% in 2090. Under the hotter and drier Canadian climate scenario, dryland soybean yields decrease 10-30% in some key locations by 2030 and decrease by 80% by 2090. Economic impact simulations follow patterns similar to the yield maps below.

Of the major crop growing areas of the Southeast, the lower Mississippi Valley and Gulf Coast areas are likely to be more negatively affected, while the northern Atlantic Coastal Plain is likely to be more positively affected.

Adaptations: Expected impacts on agricultural productivity and profitability will very likely stimulate adjustments in management strategies. Producers can switch crops or vary planting dates, patterns of water usage, crop rotations, and the amounts, timing, and application methods for fertilizers and pesticides. Analyses indicate that farmers, except those in the southern Mississippi Delta and Gulf Coast areas, will likely be able to mitigate most of the negative effects and possibly benefit from changes in CO_2 and moisture that enhance crop growth. Improvements in understanding climate and forecasting weather would enhance the ability of agricultural resource managers to deal effectively with future changes. In addition, plant breeders could respond by developing new and improved varieties to accommodate the changed climate conditions.

The southern heat wave and drought of 1998 resulted in damages in excess of $6 billion and at least 200 deaths.

Human health concerns arise from the projected increases in maximum temperatures and heat index in the region.

Improvements in risk assessment, coastal and floodplain management, linking insurance to policies for mitigating flood damage, and local mitigation planning are strategies that are likely to decrease potential costs.

Changes in Yields of Rainfed Crops
30 year Average

Hadley Model 2030   **Hadley Model 2090**

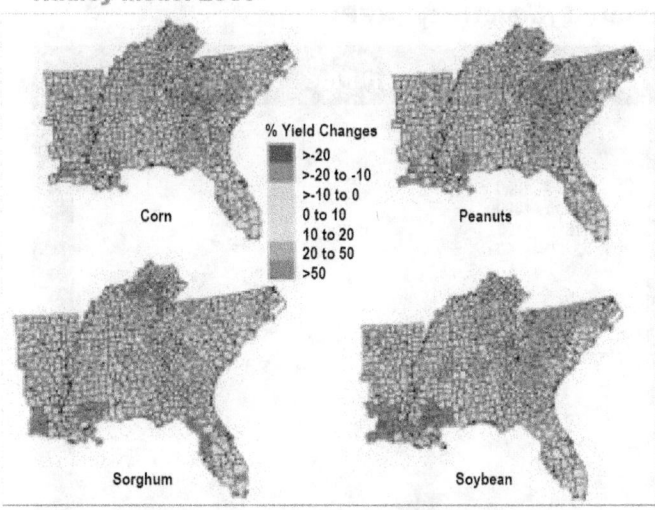
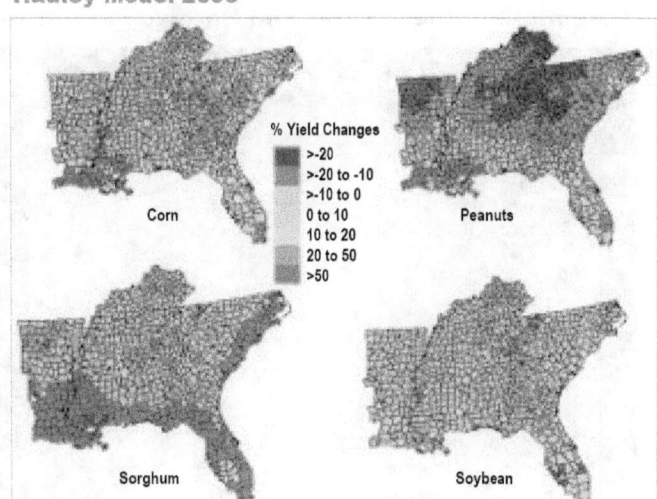

Projected changes in 30-year average rainfed yields of four major crops in the Southeast by the years 2030 and 2090 using the Hadley model scenario.

Forest Productivity Shifts

A forest process model (PnET-II) was used to evaluate the impact of the Hadley climate scenario and increasing atmospheric CO_2 on southeastern forest productivity. The model simulates an increase in the productivity of southern loblolly pine plantations of approximately 11% by 2040 and 8% by 2100; the productivity of hardwood and mixed pine hardwood forest (which represent 64% of the total forest area) would increase 22% by 2040 and 25% by 2100, compared to 1990. The model indicates that the greatest increases in productivity of both pines and hardwoods would occur in the northern half of the region.

Other VEMAP ecosystem models used with the Hadley Scenario also project increases in productivity across southern forests by 2100. However, when these models are run with the Canadian climate scenario, they simulate decreases in productivity in parts of the Southeast. Furthermore, several models that are designed to project changes in vegetation distribution as a consequence of climate change simulate a breakup of the pine-dominated forests in parts of the Southeast by the end of the 21st century under the Canadian scenario. These simulations suggest that part of the forest will possibly be replaced by savannas and grasslands due to decreased soil moisture and fire (see Ecosystems).

Adaptations: As the northern parts of the region become relatively more productive as a result of climate change and the southern parts are more negatively affected, timber harvesting could be shifted northward. Other adaptation strategies include the use of more drought-hardy strains of pine and other silvicultural and genetic improvements that could increase water use efficiency or water availability. Improved knowl-edge of the role of hurricanes, droughts, fire, El Niño-related changes in seasonal weather patterns, and other natural disturbances will be important in developing forest management regimes and increases in productivity that are sustainable over the long term. Under a hotter, drier climate, an aggressive fire management strategy could prove to be very important in this region.

Water Quality Stresses

Surface water resources in the Southeast are intensively managed with dams and channels, and almost all are affected by human activities. In some streams and lakes, water quality is either below recommended levels or nearly so. Stresses on water quality are associated with intensive agricultural practices, urban development, coastal processes, and mining activities. The impacts of these stresses are likely to be exacerbated by climate change. For example, higher temperatures reduce dis-

Potential Southern Pines and Hardwoods Net Primary Productivity (NPP)

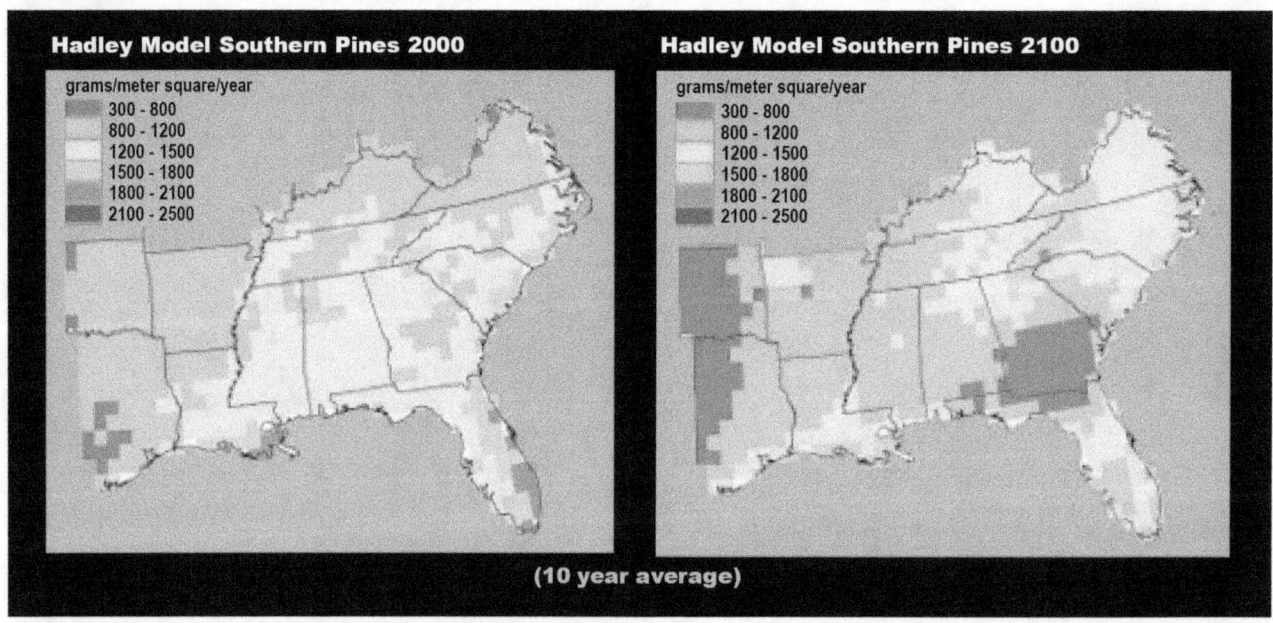

Hadley Model Southern Pines 2000

grams/meter square/year
- 300 - 800
- 800 - 1200
- 1200 - 1500
- 1500 - 1800
- 1800 - 2100
- 2100 - 2500

Hadley Model Southern Pines 2100

grams/meter square/year
- 300 - 800
- 800 - 1200
- 1200 - 1500
- 1500 - 1800
- 1800 - 2100
- 2100 - 2500

(10 year average)

solved oxygen levels in water. The 1999 flooding of eastern North Carolina offers a graphic example of how water quality can also be affected by extreme precipitation events, the frequency of which are likely to continue to increase; flood waters fouled with sewage, rotting farm animal carcasses, fuel, and chemicals swamped water treatment plants and contaminated public water supplies.

Threats to Coastal Areas

Sea-level rise is one of the more certain consequences of climate change. It has already had significant impacts on coastal areas and these impacts are very likely to increase. Between 1985 and 1995, southeastern states lost more than 32,000 acres of coastal salt marsh due to a combination of human development activities, sea-level rise, natural subsidence, and erosion. About 35 square miles of coastal land were lost each year in Louisiana alone from 1978 to 1990. Flood and erosion damage stemming from sea-level rise coupled with storm surges are very likely to increase in coastal communities. Coastal ecosystems and the services they provide to human society are

likely to be negatively affected. Projected impacts are likely to include the loss of barrier islands and wetlands that protect coastal communities and ecosystems from storm surges, reduced fisheries productivity as coastal marshes and submerged grass beds are displaced or eliminated, and saltwater intrusion into surface and ground water supplies. The extent of the ecological impacts of sea-level rise is largely dependent upon the rate of rise and the development that has occurred along the shoreline. Other threats to these ecosystems come from changes in rainfall in coastal watersheds which are likely to alter fresh water inflows into estuaries, altering salinity patterns that determine the type and distribution of coastal plant and animal communities. There are few practical options for protecting natural ecosystems as a whole from increasing temperature, changes in precipitation, or rapidly rising sea level.

As noted for other coastal regions, one possibility is the acquisition of lands contiguous to coastal wetlands to allow for their inland migration as sea level rises.

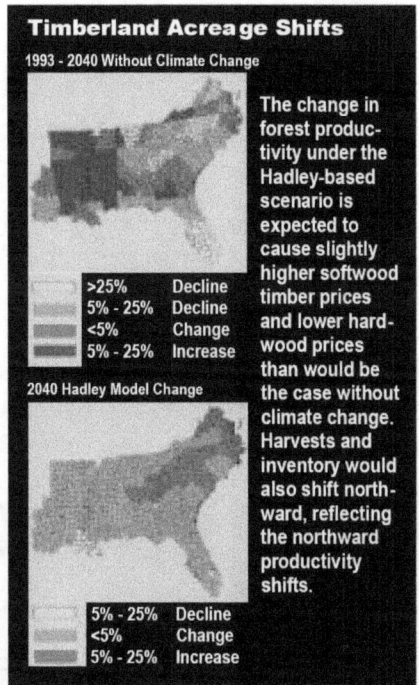

Timberland Acreage Shifts

1993 - 2040 Without Climate Change

>25%	Decline
5% - 25%	Decline
<5%	Change
5% - 25%	Increase

2040 Hadley Model Change

5% - 25%	Decline
<5%	Change
5% - 25%	Increase

The change in forest productivity under the Hadley-based scenario is expected to cause slightly higher softwood timber prices and lower hardwood prices than would be the case without climate change. Harvests and inventory would also shift northward, reflecting the northward productivity shifts.

Simulations of Net Primary Productivity (the net amount of carbon fixed by green plants over the course of a year) of southern pines and hardwoods as projected by one ecological model, PnET, using the Hadley model scenario. By 2100, PnET projects that southern hardwoods will be much more productive than pines under the climate projected by the Hadley model.

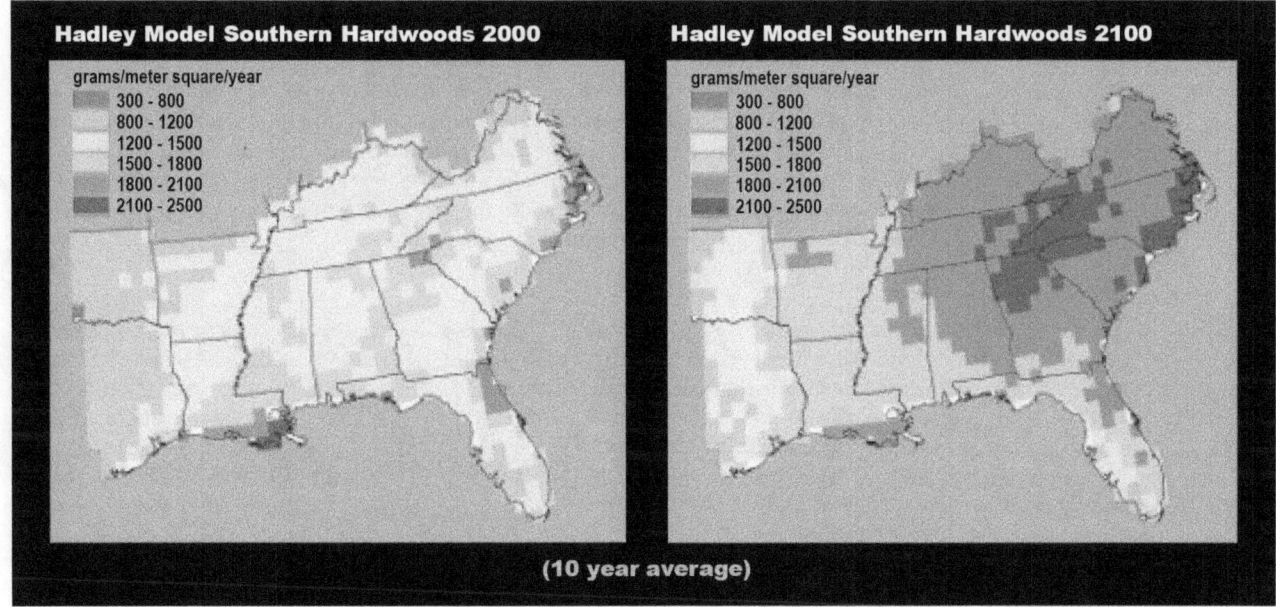

Hadley Model Southern Hardwoods 2000

grams/meter square/year
- 300 - 800
- 800 - 1200
- 1200 - 1500
- 1500 - 1800
- 1800 - 2100
- 2100 - 2500

Hadley Model Southern Hardwoods 2100

grams/meter square/year
- 300 - 800
- 800 - 1200
- 1200 - 1500
- 1500 - 1800
- 1800 - 2100
- 2100 - 2500

(10 year average)

MIDWEST

Farming, manufacturing, and forestry characterize the Midwest. The Great Lakes form the world's largest freshwater lake system, providing a major recreation area as well as a regional water transportation system with access to the Atlantic Ocean via the St. Lawrence Seaway. The region encompasses the headwaters and upper basin of the Mississippi River and most of the length of the Ohio River, both critical water sources and means of industrial transportation providing an outlet to the Gulf of Mexico. The Midwest contains some of the richest farmland in the world and produces most of the Nation's corn and soybeans. It also has important metropolitan centers, including Chicago and Detroit. Most of the largest urban areas in the region are found along the Great Lakes and major rivers. The "North Woods" are a large source of forestry products and have the advantage of being situated near the Great Lakes, providing for easy transportation.

Observed Climate Trends

Over the 20th century, the northern portion of the Midwest, including the upper Great Lakes, has warmed by almost 4°F (2°C), while the southern portion, along the Ohio River valley, has cooled by about 1°F (0.5°C). Annual precipitation has increased, with many of the changes quite substantial, including as much as 10 to 20% increases over the 20th century. Much of the precipitation has resulted from an increased rise in the number of days with heavy and very heavy precipitation events. There have been moderate to very large increases in the number of days with excessive moisture in the eastern portion of the basin.

Scenarios of Future Climate

During the 21st century, models project that temperatures will increase throughout the Midwest, and at a greater rate than has been observed in the 20th century. Even over the northern portion of the region, where warming has been the largest, an accelerated warming trend is projected for the 21st century, with temperatures increasing by 5 to 10°F (3 to 6°C). The average minimum temperature is likely to increase as much as 1 to 2°F (0.5 to 1°C) more than the maximum temperature. Precipitation is likely to continue its upward trend, at a slightly accelerated rate; 10 to 30% increases are projected across much of the region. Despite the increases in precipitation, increases in temperature and other meteorological factors are likely to lead to a substantial increase in evaporation, causing a soil moisture deficit, reduction in lake and river levels, and more drought-like conditions in much of the region. In addition, increases in the proportion of precipitation coming from heavy and extreme precipitation are very likely.

Climate Extremes Create Critical Transportation Problems

Climate extremes in the Midwest can drastically impede the highly weather-sensitive transportation systems that serve not only the region, but the entire nation. Chicago is the nation's rail hub handling much of the nation freight traffic. Barges operating on the Mississippi River system, that includes the Ohio, Illinois, and Missouri Rivers, handle a large fraction of the country's bulk commodities, such as grain and coal.

Prolonged heavy rainfall in the spring and summer of 1993 produced extensive flooding across nine states in the upper Midwest. The flood waters poured over and through many levees and inundated numerous floodplains that many of the key rail lines cross. The flood waters became an

An accelerated warming trend is projected for the 21st century, with temperatures increasing by 5 to 10°F (3 to 6°C). Precipitation is likely to continue its upward trend, at a slightly accelerated rate; 10 to 30% increases are projected across much of the region.

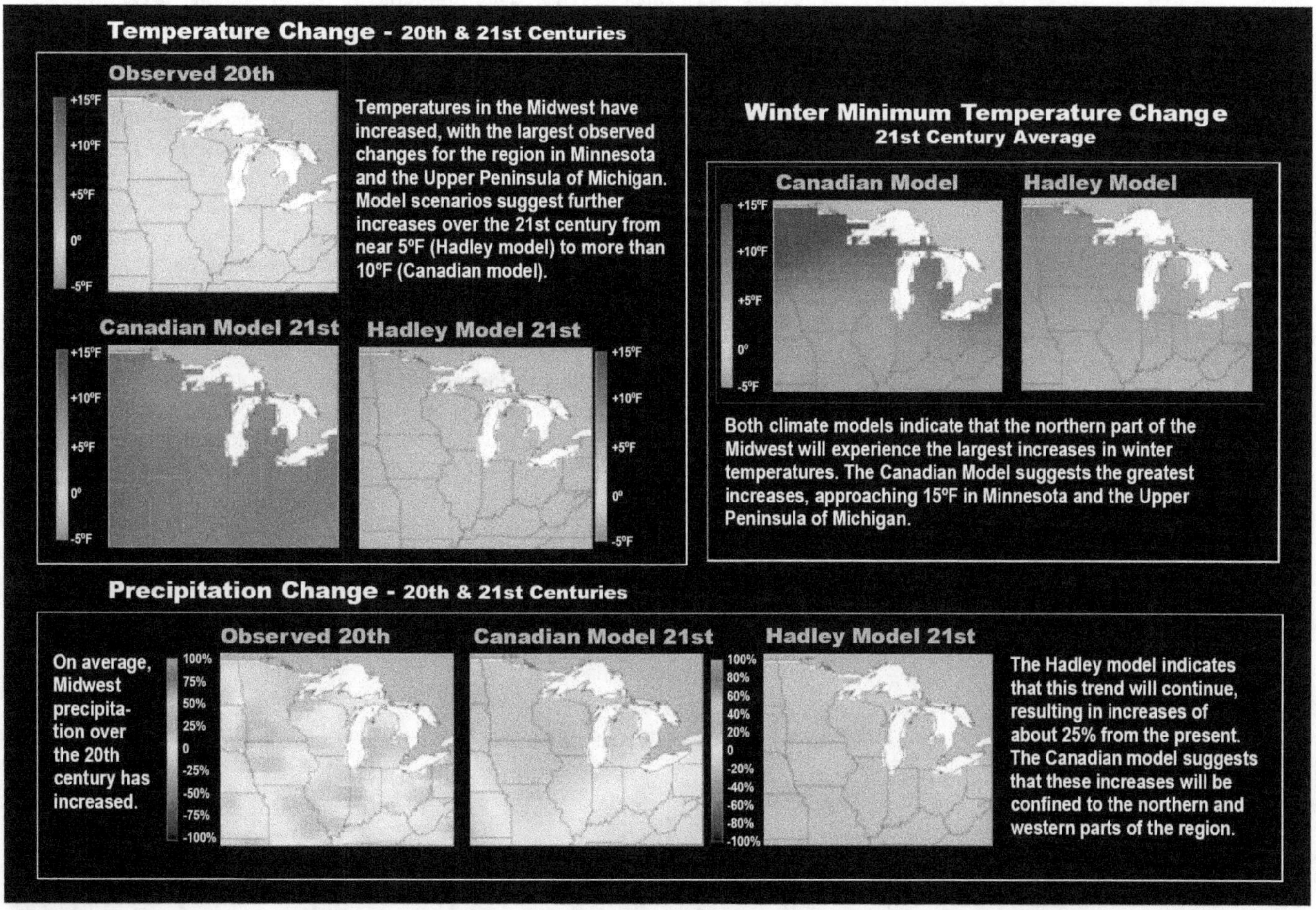

Temperature Change - 20th & 21st Centuries

Observed 20th

+15°F
+10°F
+5°F
0°
-5°F

Temperatures in the Midwest have increased, with the largest observed changes for the region in Minnesota and the Upper Peninsula of Michigan. Model scenarios suggest further increases over the 21st century from near 5°F (Hadley model) to more than 10°F (Canadian model).

Canadian Model 21st

+15°F
+10°F
+5°F
0°
-5°F

Hadley Model 21st

+15°F
+10°F
+5°F
0°
-5°F

Winter Minimum Temperature Change
21st Century Average

Canadian Model   Hadley Model

+15°F
+10°F
+5°F
0°
-5°F

Both climate models indicate that the northern part of the Midwest will experience the largest increases in winter temperatures. The Canadian Model suggests the greatest increases, approaching 15°F in Minnesota and the Upper Peninsula of Michigan.

Precipitation Change - 20th & 21st Centuries

Observed 20th   Canadian Model 21st   Hadley Model 21st

On average, Midwest precipitation over the 20th century has increased.

100%
75%
50%
25%
0
-25%
-50%
-75%
-100%

100%
80%
60%
40%
20%
0
-20%
-40%
-60%
-80%
-100%

The Hadley model indicates that this trend will continue, resulting in increases of about 25% from the present. The Canadian model suggests that these increases will be confined to the northern and western parts of the region.

absolute barrier to surface transportation in the region for more than six weeks. Train traffic had to be rerouted around the flood area, resulting in long delays and large costs to manufacturing. River barge traffic suffered a similar fate with the additional costs to shipping and manufacturing approaching $2 billion.

This came on the heels of the 1988 drought that also had a major impact on barge shipping due to low river levels, illustrating the sensitivity of transportation systems to both wet and dry climate extremes.

MIDWEST KEY ISSUES

Reduction in Lake and River Levels

Water levels, supply, quality, and water-based transportation and recreation are all climate-sensitive issues affecting the region. Despite the projected increase in precipitation, increased evaporation due to higher summer air temperatures is likely to lead to reduced levels in the Great Lakes. Of 12 models used to assess this question, 11 suggest significant decreases in lake levels while one suggests a small increase. The total range of the 11 models' projections is less than a one-foot increase to more than a five-foot decrease. A five-foot (1.5-meter) reduction would lead to a 20 to 40% reduction in outflow to the St. Lawrence Seaway. Lower lake levels cause reduced hydropower generation downstream, with reductions of up to 15% by 2050. An increase in demand for water across the region at the same time as net flows decrease is of particular concern. There is a possibility of increased national and international tension related to increased pressure for water diversions from the Lakes as demands for water increase. For smaller lakes and rivers, reduced flows are likely to cause water quality issues to become more acute. In addition, the projected increase in very heavy precipitation events will likely lead to increased flash flooding and worsen agricultural and other non-point source pollution as more frequent heavy rains wash pollutants into rivers and lakes. Lower water levels are likely to make water-based transportation more difficult with increases in the costs of navigation of 5 to 40%. Some of this increase will likely be offset as reduced ice cover extends the navigation season. Shoreline damage due to high lake levels is likely to decrease 40 to 80% due to reduced water levels.

Adaptations: A reduction in lake and river levels would require adaptations such as re-engineering of ship docks and locks for transportation and recreation. If flows decrease while demand increases, international commissions focusing on Great Lakes water issues are likely to become even more important in the future. Improved forecasts and warnings of extreme precipitation events could help reduce some related impacts.

The projected increase in very heavy precipitation events will likely worsen agricultural and other non-point source pollution as more frequent heavy rains wash pollutants into rivers and lakes.

Lower water levels are likely to make water-based transportation more difficult with increases in the costs of navigation of 5 to 40%.

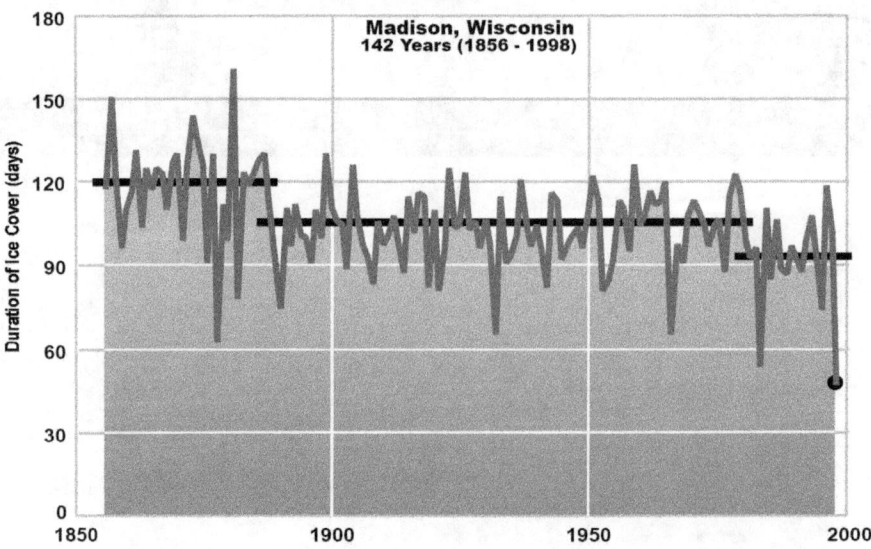

Lake Ice Duration at Lake Mendota

Madison, Wisconsin
142 Years (1856 - 1998)

Duration of Ice Cover (days)

Lake ice duration has decreased by nearly one month over the past 150 years, with a record low in the winter of 1997-98. This is consistent with observed increases in temperature.

Health and Quality of Life in Urban Areas

A reduction in extremely low temperatures and an increase in extremely high temperatures are expected. Thus,a reduced risk of life-threatening cold and an increased risk of life-threatening heat are likely to accompany warming. Reduced expenditures on snow and ice removal and fewer snow and ice related accidents and delays are likely. During the summer, however, in cities,heat-related stresses are very likely to be exacerbated by the urban heat island effect,a phenomenon in which cities remain much warmer than surrounding rural areas. This elevates nighttime temperatures,and in combination with the greater expected rise of nighttime temperatures compared to those of daytime,there will be less relief at night during heat waves. Elevated nighttime temperatures were a notable characteristic of the 1995 heat wave that resulted in over 700 deaths in Chicago. In addition,during heat waves in the Midwest, air pollutants are trapped near the surface,as atmospheric ventilation is reduced. Without strict attention to regional emissions of air pollutants,the undesirable combination of extreme heat and unhealthy air quality is likely to result. There is also a possibility of an increased risk of water-borne diseases with increases in extreme precipitation events,and increased insect- or tick-borne diseases,such as St.Louis encephalitis. Recreational activities will very likely shift as cold-season recreation such as skiing,snowmobiling,ice skating, and ice-fishing,are reduced,and warm-season recreation such as swimming,hiking,and golf, are expanded,although during mid-summer, these activities are likely to be affected by excessive heat.

Adaptations: Active responses,such as those taken by Chicago during the 1999 heat wave,are likely to help reduce the death toll due to extreme heat. Separate storm water and sewer lines and other appropriate preventative measures can help mitigate the possible increased risk of water-borne diseases.

During the summer, in cities, heat-related stresses are very likely to be exacerbated by the urban heat island effect, a phenomenon in which cities remain much warmer than surrounding rural areas.

Elevated nighttime temperatures were a notable characteristic of the 1995 heat wave that resulted in over 700 deaths in Chicago.

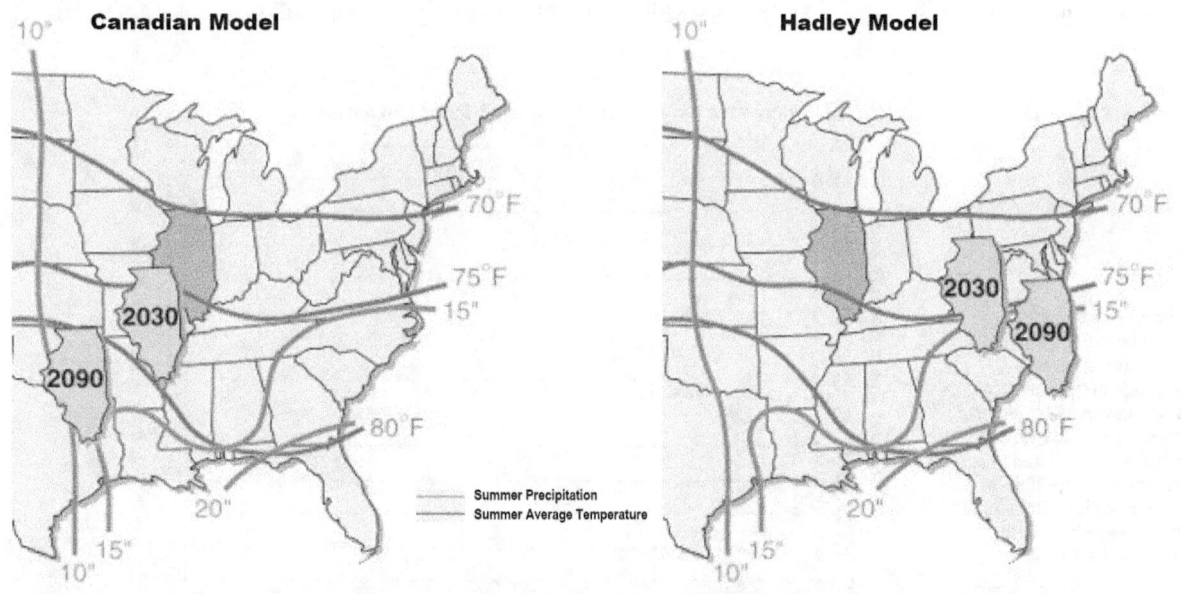

Illustration of how the summer climate of Illinois would shift under the Canadian and Hadley model scenarios. Under the Canadian scenario, the summer climate of Illinois would become more like the current climate of southern Missouri in 2030 and more like Oklahoma's current climate in 2090. The primary difference in the resulting climates of the two models relates to the amount of summer rainfall.

Agricultural Shifts

Agriculture is of vital importance to this region,the nation,and the world. It has exhibited a capacity to adapt to moderate differences in growing season climate,and it is likely that agriculture would be able to continue to adapt. With an increase in the length of the growing season, double cropping,the practice of planting a second crop after the first is harvested,is likely to become more prevalent. The CO_2 fertilization effect is likely to enhance plant growth and contribute to generally higher yields. The largest increases are projected to occur in the northern areas of the region,where crop yields are currently temperature limited. However,

yields are not likely to increase in all parts of the region. For example,in the southern portions of Indiana and Illinois,corn yields are likely to decline,with 10-20% decreases projected in some locations. Consumers are likely to pay lower prices due to generally increased yields,while most producers are likely to suffer reduced profits due to declining prices. Increased use of pesticides and herbicides are very likely to be required and to present new challenges.

Adaptations: Plant breeding programs can use skilled climate predictions to aid in breeding new varieties for the new growing conditions. Farmers can then choose varieties that are better attuned to the expected climate. It is likely that plant breeders will need to use all the tools of plant breeding,including genetic engineering,in adapting to climate change.Changing planting and harvest dates and planting densities,and using integrated pest management, conservation tillage,and new farm

technologies are additional options. There is also the potential for shifting or expanding the area where certain crops are grown if climate conditions become more favorable. Weather conditions during the growing season are the primary factor in year-to-year differences in corn and soybean yields. Droughts and floods result in large yield reductions;severe droughts,like the drought of 1988, cause yield reductions of over 30%. Reliable seasonal forecasts are likely to help farmers adjust their practices from year to year to respond to such events.

Farm flooded by Mississippi river in 1993.

The relationship between Midwest soybean yield and precipitation is shown here. Soybean yields in thousands of bushels are shown as the differences from the average yield in recent decades. Precipitation is the difference from the 1961-90 average precipitation. Note that lower yields result from both extreme wet and extreme dry conditions.

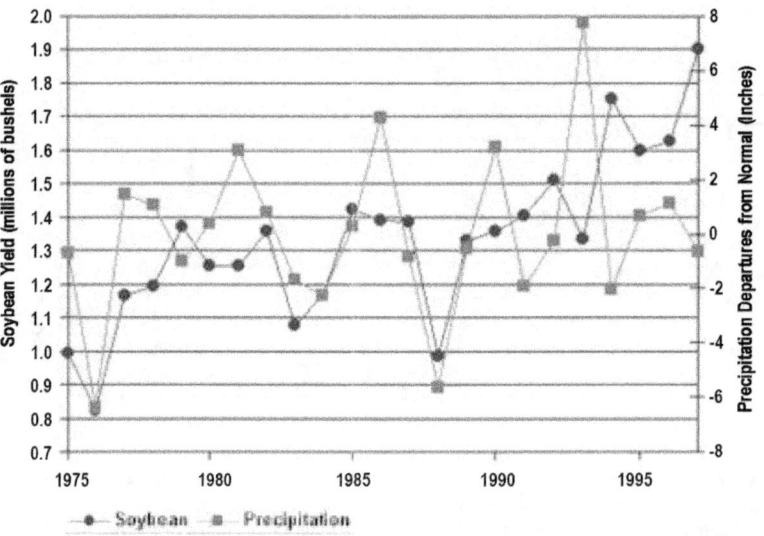

Midwest Soybean Yield and Precipitation

Changes in Semi-natural and Natural Ecosystems

The upper Midwest has a unique combination of soil and climate that allows for abundant coniferous tree growth. Higher temperatures and increased evaporation will likely reduce boreal forest acreage,and make current forestlands more susceptible to pests and diseases. It is likely that the southern transition zone of the boreal forest will be susceptible to expansion of temperate forests, which in turn will have to compete with other land use pressures. However, warmer weather (coupled with beneficial effects of increased CO_2),are likely to lead to an increase in tree growth rates on marginal forestlands that are currently temperature-limited. Most climate models indicate that higher air temperatures will cause greater evaporation and hence reduced soil moisture, a situation conducive to forest fires. As the 21st century progresses,there will be an increased likelihood of

greater environmental stress on both deciduous and coniferous trees,making them susceptible to disease and pest infestation,likely resulting in increased tree mortality.

As water temperatures in lakes increase,major changes in freshwater ecosystems will very likely occur, such as a shift from cold water fish species,such as trout,to warmer water species,such as bass and catfish. Warmer water is also likely to create an environment more susceptible to invasions by non-native species. Runoff of excess nutrients (such as nitrogen and phosphorus from fertilizer) into lakes and rivers is likely to increase due to the increase in heavy precipitation events. This, coupled with warmer lake temperatures,is likely to stimulate the growth of algae,depleting the water of oxygen to the detriment of other living things. Declining lake levels are likely to cause large impacts to the current distribution of wetlands. There is some chance that some wet-

lands could gradually migrate,but in areas where their migration is limited by the topography, they would disappear. Changes in bird populations and other native wildlife have already been linked to increasing temperatures and more changes are likely in the future. Wildlife populations are particularly susceptible to climate extremes due to the effects of drought on their food sources.

Runoff of excess nutrients (such as nitrogen and phosphorus from fertilizer) into lakes and rivers is likely to increase due to the increase in heavy precipitation events.

Projected Midwest Daily Precipitation
21st Century

Canadian Model

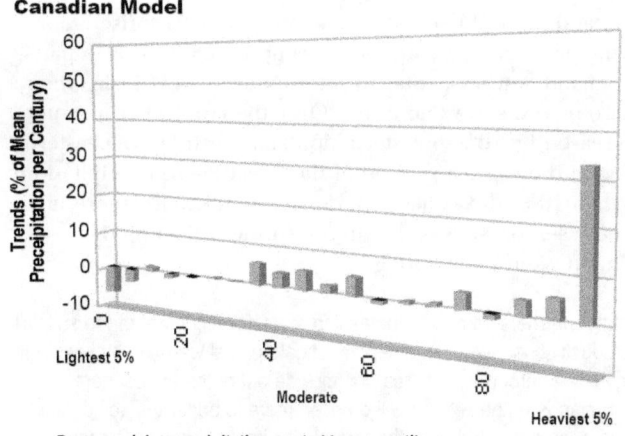

Hadley Model

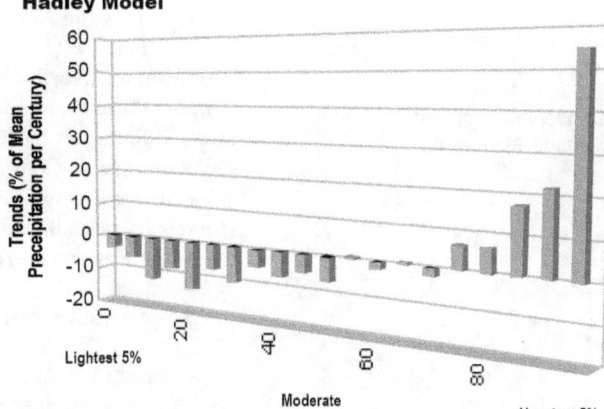

Days receiving precipitation, sorted by percentile.

Annual trends in daily precipitation by percentile for the Canadian and Hadley model scenarios for the 21st century. Notice the largest trend is in the heaviest daily precipitation amount for both model simulations, indicating that most of the projected increase in annual precipitation will be due to an increase in precipitation on days already receiving large amounts.

GREAT PLAINS

While farming and ranching are still the primary land uses of the Great Plains, urban areas provide housing and jobs for two-thirds of the region's people. Native ecosystems and agricultural fields intermingle with small rural communities and the expanding metropolitan centers. The region produces much of the nation's grain, meat, and fiber, including over 60% of the wheat, 87% of the sorghum, and 36% of the cotton. The region is home to over 60% of the nation's livestock, including both grazing and grain-fed-cattle operations. Recreation, wildlife habitat, and water resources are also found on the landscapes of the Great Plains. The Prairie Pothole region provides important habitat for migratory waterfowl. Surface water in rivers, streams, and lakes, and groundwater in aquifers provide water for urban, industrial, and agricultural uses, as well as riparian and aquatic ecosystems. Soil organic matter is a major resource of the Great Plains as it provides improved soil water retention, soil fertility, and long-term storage of carbon.

Climate determines many aspects of life on the Plains. For agriculture, weather determines the planting and harvesting dates for crops, livestock grazing and breeding seasons, and water availability. The high natural variability of climate is a characterizing feature of the region. Farmers and ranchers have survived by being adaptive and incorporating new technologies to buffer their production against the variable climate. For urban centers, the availability of water often constrains urban and industrial development. Dams, diversions, channels, and groundwater pumping have influenced nearly all freshwater ecosystems in the region. Ongoing social and economic changes in the Great Plains will continue to increase demands on the region's land and water resources and challenge its physical and social infrastructure. Climate change will present additional challenges to long-term planning for new infrastructure and the sustainable use of land and water.

Observed Climate Trends

Across the Northern and Central Great Plains, temperatures have risen more than 2°F (1°C) in the past century, with increases up to 5.5°F (3°C) in parts of Montana, North Dakota, and South Dakota. In the southern Great Plains, the 20th century temperature record shows no trend. Over the last 100 years, annual precipitation has decreased by 10% in eastern Montana, North Dakota, eastern Wyoming, and Colorado. In the eastern portion of the Great Plains, precipitation has increased by more than 10%. Texas has experienced significantly more high intensity rainfall. The snow season ends earlier in the spring, reflecting the greater seasonal warming in winter and spring.

KEY ISSUES

- Alteration in Timing and Amount of Water

- Changes in Climate Extremes

- Invasive Species Effects on Economy and Ecology

- Stress on Human Communities

- Conservation of Soil Organic Matter

Increasing Soil Carbon Helps Buffer Against Climate Change Impacts

Martin Kleinschmit, a farmer and rancher in Bow Valley, Nebraska, says that farmers have a lot at risk as global climate heats up, but they also have a lot to gain by participating in the solution to climate change. By conserving soil organic matter, farmers can improve soil health and productivity as well as capture and store (sequester) carbon in the extensive crop and rangelands of the Great Plains. The higher temperatures and greater numbers of droughts and floods projected for the region could threaten crops, raise production expenses, and increase the risk of failure. To protect our food supply, healthy soils able to withstand erratic weather patterns are needed.

Increasing the carbon content of the soil will help to mitigate global warming by keeping carbon dioxide out of the atmosphere, but it will do even more to buffer the soil against the threats of climate change. Presently, most US farmland has only half or less of its historical level of organic matter. Soil scientists have established that a 6-inch (15 cm) block of soil with 1 to 2% organic matter can hold only about one inch (2.5 cm) of rain before it runs out the bottom. With 4

Scenarios of Future Climate

Climate model scenarios project that temperatures will continue to rise throughout the region, with the largest increases in the western parts of the Plains. The Canadian model projects greater increases throughout the region than the Hadley model. The climate model projections, as well as other tools utilized in analyzing impacts, include a greater number of heat events – three days in a row above 90°F – a major cause of heat stress for people and livestock. Seasonally, more warming is expected in winter and spring than in summer and fall. Precipitation generally increases in the region in the Hadley model, and in the northern parts in the Canadian model. Precipitation decreases in the lee of the Rocky Mountains in both models. This is accentuated in the Canadian model, with decreases of up to 25% in an area centered on the Oklahoma panhandle and covering northern Texas, eastern Colorado, and western Kansas. Smaller decreases are seen in the Hadley model in a band from northern Texas through Montana. Although precipitation increases are projected for parts of the Great Plains, increased evaporation due to rising air temperatures are projected to surpass these increases, resulting in net soil moisture declines for large parts of the region.

Seasonally, more warming is expected in winter and spring than in summer and fall. Precipitation generally increases in the region in the Hadley model, and in the northern parts in the Canadian model. Precipitation decreases in the lee of the Rocky Mountains in both models.

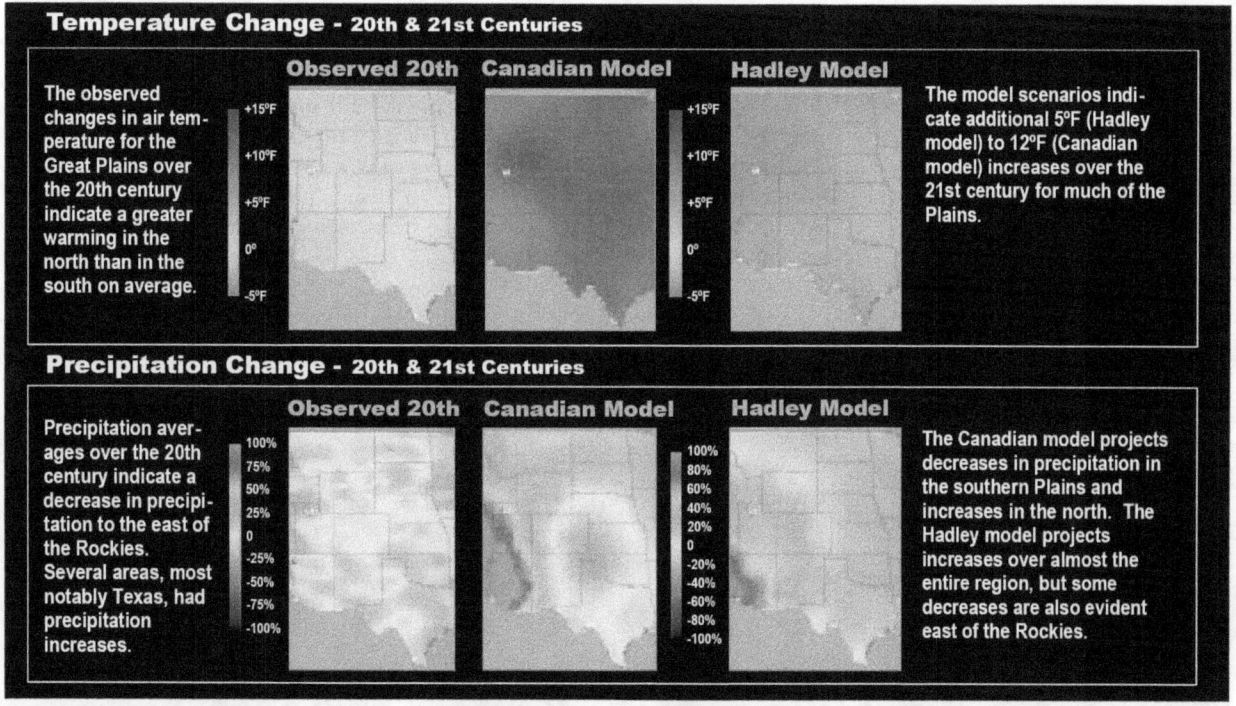

Temperature Change - 20th & 21st Centuries

The observed changes in air temperature for the Great Plains over the 20th century indicate a greater warming in the north than in the south on average.

Observed 20th | Canadian Model | Hadley Model

+15°F / +10°F / +5°F / 0° / -5°F

The model scenarios indicate additional 5°F (Hadley model) to 12°F (Canadian model) increases over the 21st century for much of the Plains.

Precipitation Change - 20th & 21st Centuries

Precipitation averages over the 20th century indicate a decrease in precipitation to the east of the Rockies. Several areas, most notably Texas, had precipitation increases.

Observed 20th | Canadian Model | Hadley Model

100% / 75% / 50% / 25% / 0 / -25% / -50% / -75% / -100%

100% / 80% / 60% / 40% / 20% / 0 / -20% / -40% / -60% / -80% / -100%

The Canadian model projects decreases in precipitation in the southern Plains and increases in the north. The Hadley model projects increases over almost the entire region, but some decreases are also evident east of the Rockies.

to 5% organic matter, that same soil can hold 4-6 inches (10 to 15 cm) of rain before it leaves the root zone and takes with it the water-soluble nutrients. Increasing soil organic matter also reduces the risks of flooding and erosion, and retains moisture longer so plants have access to it during periods of dry weather. Soil organic matter lessens the need for (and expense of) irrigation, reduces ground water pollution, and reduces the amount of run-off, lessening

the threat of stream pollution. It also lowers the cost of fertilization since nutrients not lost to erosion and leaching need not be replaced. Agricultural incentives that encourage net carbon sequestration in soil provide an opportunity to promote food security in a changing climate and reduce the threat of climate change at the same time.

GREAT PLAINS KEY ISSUES

Alteration in Timing and Amount of Water

Water supply, demand, allocation, storage, and quality are all climate-sensitive issues affecting the regional economy. Farming and ranching use over 50% of the region's water resources. Ground-water pumping for irrigation has depleted aquifers in portions of the Great Plains by withdrawing water much faster than it can be recharged. Under today's irrigation demands, water table levels are thus dropping in parts of the southern Great Plains. The projected climate-induced changes in water resources are likely to exacerbate the current competition for water among the agricultural sector, natural ecosystems, and urban, industrial, and recreational users.

Adaptations: It is possible that current strategies to deal with drought, water shortages, extreme weather, and variability could help the region cope with future climate change impacts. These strategies include switching to crops that use less water, retiring marginal lands, adopting conservation tillage, and enhanced watershed storage capacity and groundwater recharge activities. Water availability for crops could possibly be improved using new and existing technologies for crop residue management, wind breaks, mulches, soil carbon management, tillage practices, precision agriculture, and more efficient methods of water application. While these strategies would improve water use efficiency, adaptive strategies would also need to include maintaining water quality. Flexible policies and institutions would help to adapt to unanticipated hydrologic changes.

The projected climate-induced changes in water resources are likely to exacerbate the current competition for water among the agricultural sector, natural ecosystems, and urban, industrial, and recreational users.

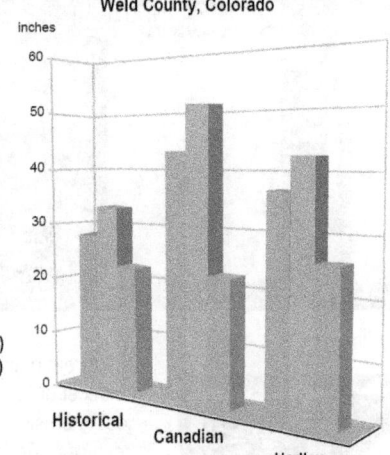

Grass Alfalfa Corn

Historical (1981-1990 average)
Canadian (2090-2100 average)
Hadley (2090-2100 average)

Lack of soil moisture can greatly reduce yield of crops and forage. Under both climate scenarios, the consumptive demand for water on grass pasture increases more than 50% while the water needs for irrigated corn change little. Perennial crops such as alfalfa experience an increase in consumptive demand for water; the size of the increase depends on the climate scenario.

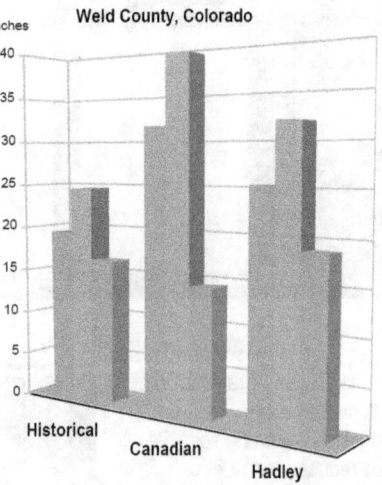

Water supplies for agriculture have been decreasing in many parts of the Great Plains, partly due to increases in urban uses. Irrigation water needs for grass and alfalfa are projected to increase under both climate scenarios while those for corn rise slightly. The changes in irrigation needs reflect the seasonal shift in precipitation that favors corn more than perennial crops such as grass, hay, and alfalfa.

Changes in Climate Extremes

The potential for new patterns in climate extremes raises questions about the ability of current coping strategies to deal with future impacts.

Extreme climate and weather events have major effects on urban and rural lives. The April 1997 flood put approximately 90% of Grand Forks,North Dakota under water and caused over $1 billion in damages. A short-term heat wave in July 1995 caused the deaths of over 4,000 feedlot cattle in Missouri. The severe drought from Fall 1995 through Summer 1996 in the agricultural regions of the southern Great Plains resulted in about $5 billion in damages. There is some chance that the projected increase in drought tendency in the Sand Hills of the Great Plains will result in expansion or shifting of sand dunes if vegetation cover is not maintained. The potential for new patterns in climate extremes raises questions about the ability of current coping strategies to deal with future impacts.

Adaptations: Better access to more accurate and timely information about near-term weather including extreme events,and longer-term forecasts could help reduce risk and uncertainty in decision making. For example,heat stress events are projected to occur more often in the central and southern Great Plains in the future. This information can help intensive-livestock operators weigh strategic decisions about investments in cooling systems. Real-time weather information can prepare them to implement an immediate response to cool their animals.

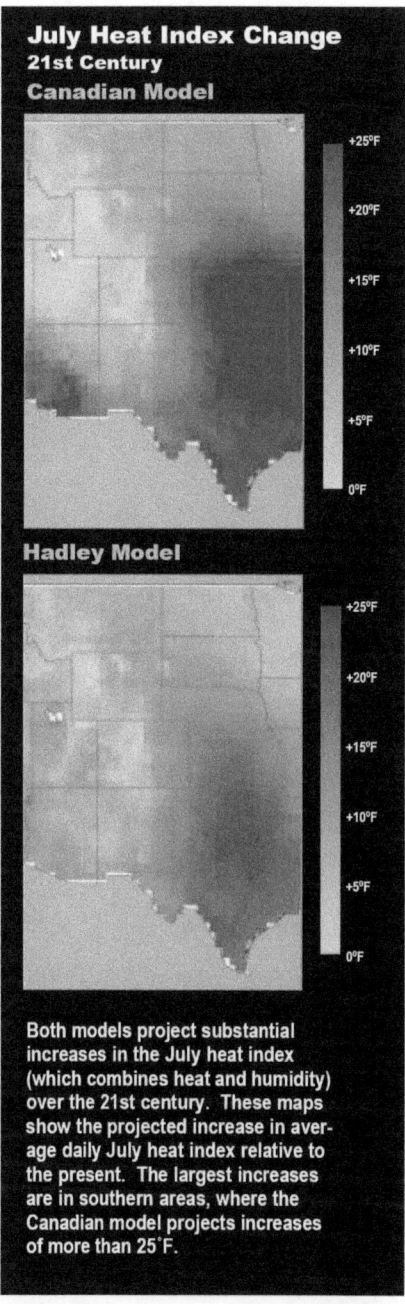

July Heat Index Change
21st Century
Canadian Model

+25°F
+20°F
+15°F
+10°F
+5°F
0°F

Hadley Model

+25°F
+20°F
+15°F
+10°F
+5°F
0°F

Both models project substantial increases in the July heat index (which combines heat and humidity) over the 21st century. These maps show the projected increase in average daily July heat index relative to the present. The largest increases are in southern areas, where the Canadian model projects increases of more than 25°F.

Palmer Drought Severity Index Change
21st Century

Canadian Model   **Hadley Model**

>10
8
6
4
2
0
-2
-4
-6
-8
-10

The Palmer Drought Severity Index (PDSI) is a commonly used measure of drought severity taking into account differences in temperature, precipitation, and capacity of soils to hold water. These maps show projected changes in the PDSI over the 21st century, based on the Canadian and Hadley climate scenarios. A PDSI of –4 indicates extreme drought conditions. The most intense droughts are in the –6 to –10 range, similar to the major droughts of the 1930s. By the end of the century, the Canadian scenario projects that extreme drought will be a common occurrence over much of the Great Plains, while the Hadley model projects much more moderate drought conditions.

GREAT PLAINS KEY ISSUES

Invasive Species Effects on Economy and Ecology

The native grasslands, shrublands, forests, and riparian ecosystems of the Great Plains are home to a variety of plants and animals. Nearly 60% of the bird species that breed in the US do so in the Great Plains. Agriculture and urban development have disrupted these native ecosystems, and invasive species are currently a serious challenge in both native ecosystems and agricultural systems. For example, leafy spurge currently reduces grazing capacity on grasslands. Field bindweed lowers crop production in Kansas by $40 million a year. Projected climate change is likely to alter the current biodiversity. A possible migration of invasive species across the Great Plains is a concern to stakeholders because the rapid rate of climate change is likely to be disadvantageous to native species. The exact social costs will depend upon the particular invasive species or type of change in biodiversity.

Adaptations: Effective coping strategies would help provide plants and animals with habitats for adaptation such as maintaining a diversity of vegetation types and connectivity between the types. Preserving intact riparian areas, wetlands, and natural areas is likely to slow or reduce future invasions and is beneficial even in the absence of climate change.

Stress on Human Communities

Rural communities, already stressed by their declining populations and shrinking economic base, are dependent on the competitive advantage of their agricultural products in domestic and foreign markets. Large corporate enterprises, the result of agribusiness modernization and consolidation, have greater resources and technology with which to buffer themselves against both economic and climatic variability. Thus, a changing climate is an additional stress that disproportionately impacts family farmers and ranchers. In urban

A possible migration of invasive species across the Great Plains is a concern to stakeholders because the rapid rate of climate change is likely to be disadvantageous to native species.

Spotted knapweed has infested over 5 million acres in Montana and is a threat to pristine natural areas such as the Grand Teton National Park in Wyoming.

Leafy spurge currently reduces grazing capacity, plant diversity, and wildlife habitat on grasslands.

Yellow starthistle, an annual herb up to 3 feet tall, currently infests over 9 million acres of rangeland in the western US, with nearly 8 million acres in California alone. This invasive species spreads as a contaminant in agricultural seeds.

Distribution of Leafy Spurge in the Western US
By County - 1996

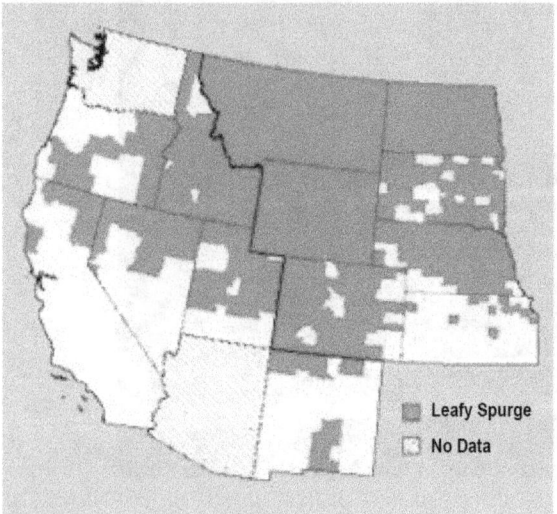

Leafy Spurge

No Data

Leafy spurge, a federally-designated noxious weed, currently can be found in 8 of the 10 Great Plains states. First recorded in 1827, leafy spurge spread from Massachusetts to North Dakota in 80 years. This deep rooted, perennial plant forms dense stands that aggressively crowd out most other vegetation, causing a loss of plant diversity, reduction of forage, and loss of wildlife habitat. Leafy spurge can spread rapidly through seed dispersal as well as being carried by birds and other animals. It can expel its seeds to distances of 15 feet.

communities,the impacts of floods, heat waves,and other climate events is a crucial emerging issue with significant economic implications.

Many poorer people can not afford air conditioning,insulation,substantial housing,and other means of coping with climate extremes.

Therefore,climate change impacts will vary significantly by social and economic status.

Adaptations:Diversification within enterprises and rural communities could help to reduce risk and cope with the additional stress of climate change. Community-level dialogue is vital in identifying information needed by managers and in assessing policy options for climate change.

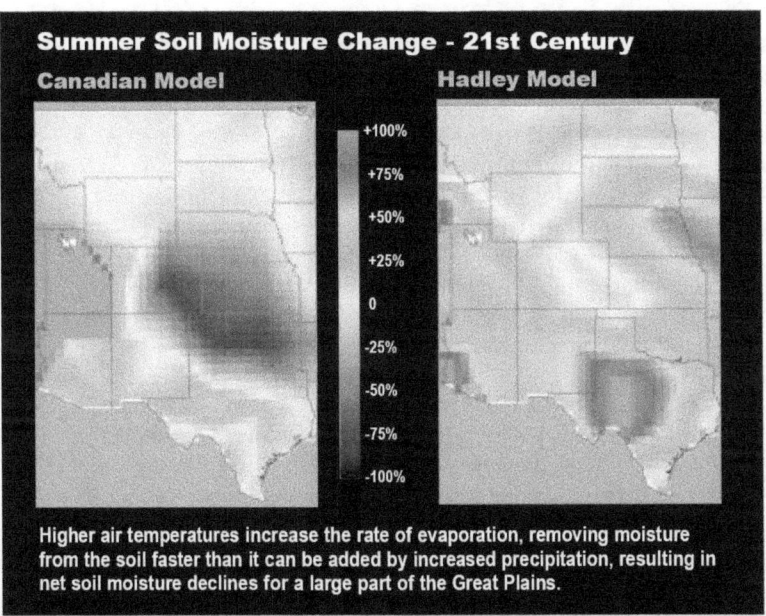

Summer Soil Moisture Change - 21st Century

Canadian Model

Hadley Model

+100%
+75%
+50%
+25%
0
-25%
-50%
-75%
-100%

Higher air temperatures increase the rate of evaporation, removing moisture from the soil faster than it can be added by increased precipitation, resulting in net soil moisture declines for a large part of the Great Plains.

Soil carbon is vital for retaining water and nutrients. The amount of carbon stored in the soil is strongly influenced by past and present land management practices and weather patterns. Overall, soil carbon is projected to decline in response to higher temperatures in both climate scenarios. In some areas, climate changes reduce the decomposition of soil organic matter, resulting in increased soil carbon.

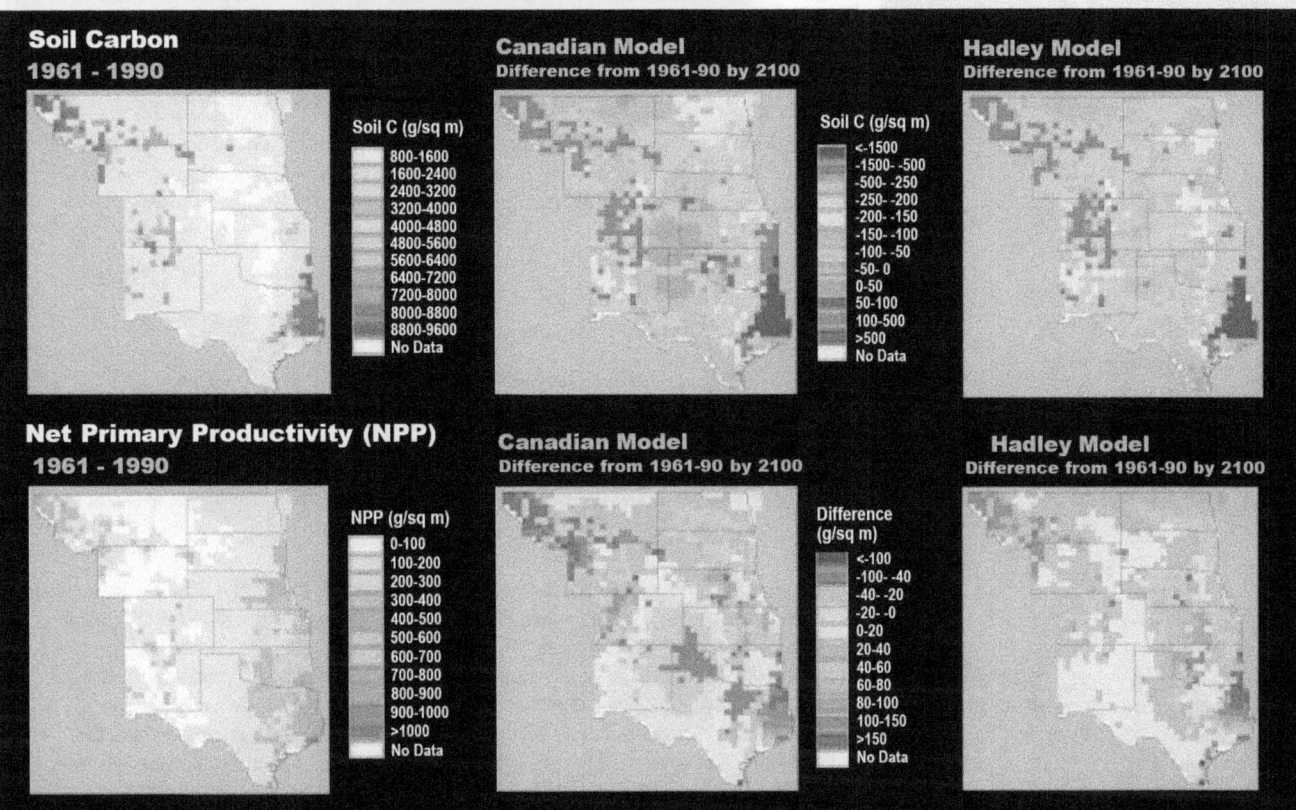

Soil Carbon
1961 - 1990

Canadian Model
Difference from 1961-90 by 2100

Hadley Model
Difference from 1961-90 by 2100

Soil C (g/sq m)

800-1600
1600-2400
2400-3200
3200-4000
4000-4800
4800-5600
5600-6400
6400-7200
7200-8000
8000-8800
8800-9600
No Data

Soil C (g/sq m)

<-1500
-1500- -500
-500- -250
-250- -200
-200- -150
-150- -100
-100- -50
-50- 0
0-50
50-100
100-500
>500
No Data

Net Primary Productivity (NPP)
1961 - 1990

Canadian Model
Difference from 1961-90 by 2100

Hadley Model
Difference from 1961-90 by 2100

NPP (g/sq m)

0-100
100-200
200-300
300-400
400-500
500-600
600-700
700-800
800-900
900-1000
>1000
No Data

Difference (g/sq m)

<-100
-100- -40
-40- -20
-20- -0
0-20
20-40
40-60
60-80
80-100
100-150
>150
No Data

The productivity of the Great Plains increases from west to east and from north to south, following the precipitation and the temperature gradients. Land uses are strongly influenced by productivity. Both climate scenarios increase the moisture stress in the central parts of the Great Plains and productivity declines in this region.

WEST

The West has a variable climate, diverse topography and ecosystems, an increasing human population, and a rapidly growing and changing economy. Western landscapes range from the coastal areas of California to the deserts of the Southwest to the alpine meadows of the Rocky and Sierra Nevada Mountains. Since 1950, the region's population has quadrupled, with most people now living in urban areas. Thus, once predominantly rural states are now among the most urban in the country. The economy of the West has been transformed from one dominated by agriculture and resource extraction to one dominated by government, manufacturing, and services. National parks attract tourists from around the world. The region has a slightly greater share of its economy in sectors that are sensitive to climate than the nation as a whole; these include agriculture, mining, construction, and tourism, which currently represent one-eighth of the region's economy.

As a result of population growth and development, the region faces multiple stresses. Among these are air quality problems, urbanization, and wildfires. Perhaps the greatest challenge, however, is water, which is typically consumed far from where it originates. Competition for water among agricultural, urban, power consumption, recreational, environmental, and other uses is intense, with water supplies already oversubscribed in many areas.

Observed Climate Trends

The climate of the West varies strongly across the region and over time. Historically, the region has experienced exceptionally wet and dry periods. During the 20th century, temperatures in the West have risen 2-5°F (1-3°C). The region has generally had increases in precipitation, with increases in some areas greater than 50%. However, a few areas, such as Arizona, have become drier and experienced more droughts. The length of the snow season decreased by 16 days from 1951 to 1996 in California and Nevada. Extreme precipitation events have increased.

The Oakland Fire and Response

Climate change is likely to increase fire frequency in the West. However, as shown by the response to the Oakland fire, there is substantial potential to reduce the risk of urban fires. On October 20, 1991, a small brush fire started in the hills above Oakland, California. Fire-conducive conditions, including high winds, unseasonably high temperatures, large stores of fuel in the form of dead plant parts, housing developments with dense and flamable vegetation, and low humidity, enabled the fire to spread rapidly. Before the fire was brought under control, it covered 1,600 acres, killed 25 people, consumed 3,229 structures and damaged another 2,992, and caused an estimated $2 billion in damage.

In response to this and other recent severe wildfires, the State developed the California Fire Plan to address pre-fire management prescriptions and improved response capabilities. Fire-prevention methods include fire-resistant construction standards for roofing and other materials, changes in zoning, and hazard reduction near structures such as vegetation clearing and management. Improved response capabilities

Scenarios of Future Climate

The two models used in this Assessment project annual average temperature increases from 3 to over 4°F (2°C) by the 2030s and 8-11°F (4.5-6°C) by the 2090s. The models project increased precipitation during winter, especially over California, where runoff is projected to double by the 2090s. In these climate scenarios, some areas of the Rocky Mountains are projected to get drier. Both models project more extreme wet and dry years. Due to uncertainties about regional precipitation, the possibility of a drier climate was also considered.

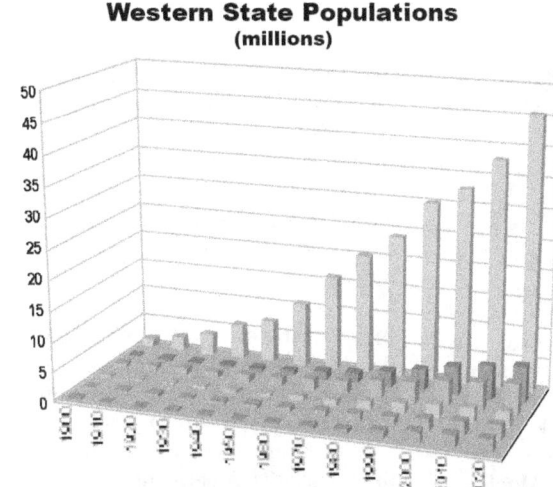

Western State Populations
(millions)

- Nevada
- New Mexico
- Utah
- Colorado
- Arizona
- California

A defining feature of the West is rapid population growth. Much of this growth has occurred in cities, and once predominantly rural states are now urbanized. Growth in urban areas is expected to continue for at least several decades. In the middle growth scenario the total population of the West is projected to grow from 48 million in 1999 to between 60 and 74 million in 2025.

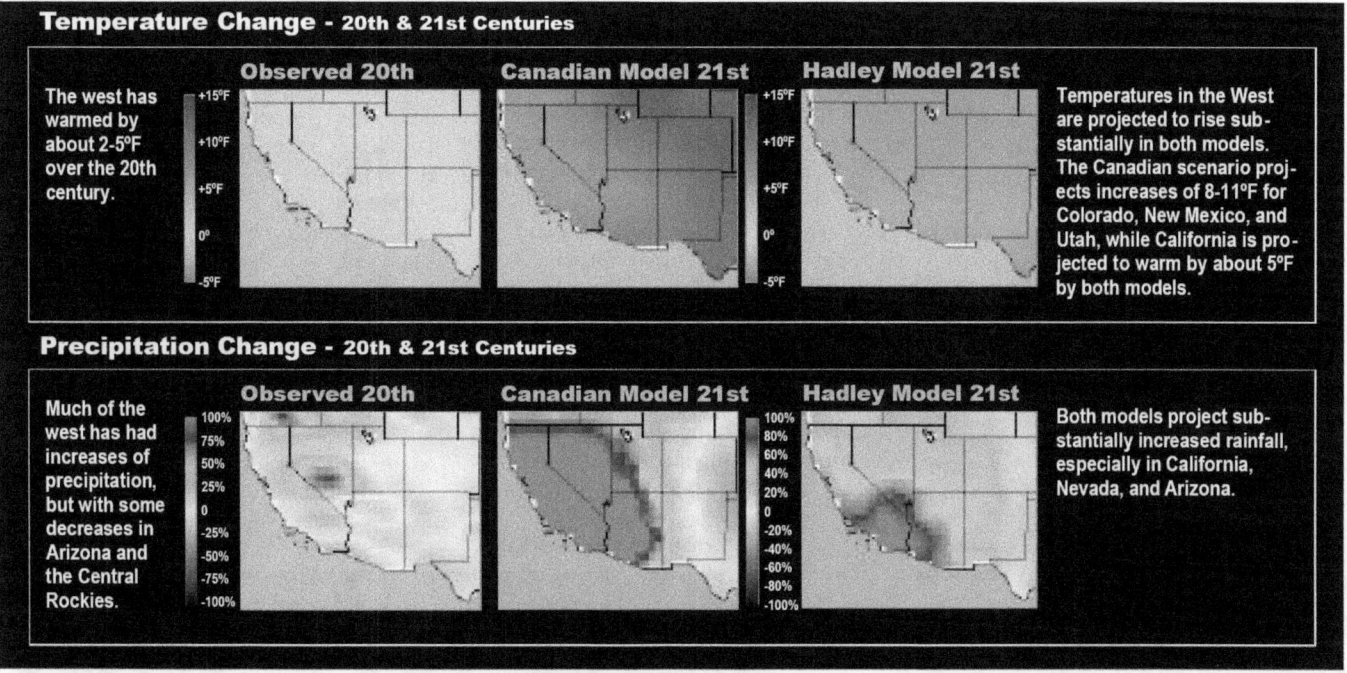

Temperature Change - 20th & 21st Centuries

Observed 20th | **Canadian Model 21st** | **Hadley Model 21st**

The west has warmed by about 2-5°F over the 20th century.

Temperatures in the West are projected to rise substantially in both models. The Canadian scenario projects increases of 8-11°F for Colorado, New Mexico, and Utah, while California is projected to warm by about 5°F by both models.

Precipitation Change - 20th & 21st Centuries

Observed 20th | **Canadian Model 21st** | **Hadley Model 21st**

Much of the west has had increases of precipitation, but with some decreases in Arizona and the Central Rockies.

Both models project substantially increased rainfall, especially in California, Nevada, and Arizona.

include better neighborhood access for firefighting equipment and adequate nearby water supplies. In addition, very high fire hazard severity zones were identified. The State requires construction ordinances for structures in these zones, including roof specifications and a minimum vegetation clearance around structures. Much of the area burned in the Oakland Hills fire has since been rebuilt according to these standards.

Studies have found that fire protection programs that included prevention elements reduced the losses from wildfires by anywhere from 50 to 80%. Results indicate that prevention strategies can aid in fuels management, control fire behavior, reduce the physical impact of fire on natural resources, improve forest health, and reduce the cost and losses due to wildfires.

WEST KEY ISSUES

Changes in Water Resources

The West's water resources are sensitive to climate change. The semiarid West is dependent upon a vast system of engineered water storage and transport, such as along the Colorado River, and is governed by complex water rights laws. Much of the water supply comes from snowmelt, and higher temperatures will very likely reduce the snowpack and alter the amount and timing of peak flows. In some places, it is likely that current reservoir systems will be inadequate to control earlier spring runoff and maintain supplies for the summer, but more research is necessary to identify which systems are most vulnerable. It is also possible that demand will increase.

In a wetter climate, the potential for flooding will increase when precipitation comes in more intense events or where total precipitation increases substantially. It is possible that more precipitation would also create additional water supplies, reduce demand, and ease competition among competing uses. Greater runoff would likely increase hydropower production and ease some water quality problems, although it is also possible that there would be more non-point source pollution.

In contrast, a drier climate is likely to decrease supplies and increase demand for such uses as agriculture, urban needs, and power production, thus making water supplies much tighter. Native Americans, among others, are exercising their rights to water, and may do so to a greater extent, further tightening supplies.

Adaptations: Improved technology, planting of less water-demanding crops, pricing water at replacement cost, and other conservation efforts can help reduce demand. Flexibility to transfer water across basins and water users, and to integrate the use of surface and groundwater, can also serve as adaptation strategies for water managers under conditions of scarcity. Environmental and cost constraints will be an important consideration in building additional flood control or storage facilities.

Changes in Natural Ecosystems

Under the Hadley and Canadian scenarios, vegetation models suggest an increase in plant growth, a reduction in desert areas, and a shift toward more woodlands and forests in many parts of the West. However, a less positive CO_2 fertilization effect than assumed in the models, increase in fires, and persistence of other stresses such as air pollution, are important sources of uncertainty. It is possible that continued increases in temperature and leveling off of the CO_2 fertilization effect would result in an eventual decline in forest productivity. A drier climate would also likely reduce forest productivity.

The diverse topography coupled with landscape fragmentation and other development pressures in the West will likely make it difficult for many species to adapt to climate change by migrating. It is likely that some ecosystems, such as alpine ecosystems, will disappear entirely from some places in the region. On the other hand it is possible that mountains may enable some species to adapt by permitting their migration to higher elevations.

Relative Consumptive Water Use in the West

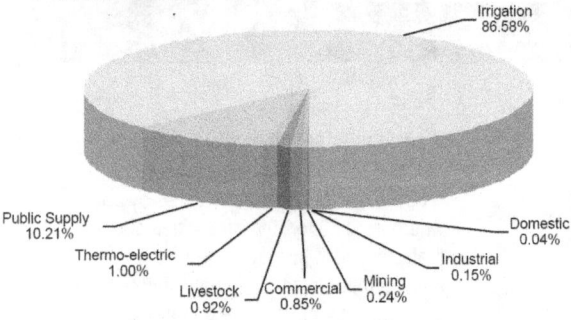

In 1995, 87% of the water used in the West was for irrigation. However, water use for irrigation has declined slightly since 1980, while municipal uses have grown.

Ecosystem Models

Climate change is projected to cause major changes in vegetation distribution during the 21st century. Overall, the model scenarios project increases in grasslands, woodlands, and forests in the West, and a loss of desert vegetation. The far left map shows potential vegetation types in the West (the vegetation that would naturally flourish in the absence of human activity given today's climate), while the two maps on the right show model-projected scenarios for future vegetation shifts in the face of climate change.

Non-native invasive species have already stressed many Western ecosystems and are likely to make adaptation to climate change much more difficult for native species. Climate change is also likely to increase fire frequency. As long as year-to-year variation in precipitation remains high, fire risk is likely to increase whether the region gets wetter or drier. This is because fuel loads tend to increase in wet years as a result of increased plant productivity and are consumed by fire in dry years. In addition, rising sea levels will threaten many coastal wetlands, such as those in the San Francisco Bay area, and the diversity of species they support.

Adaptations: Devising strategies to reduce negative climate change impacts on natural ecosystems and biodiversity is particularly challenging. Improved management of urban development can help reduce habitat fragmentation. The creation of migration corridors to help some species migrate to more suitable locations has been suggested, but its effectiveness is not known. Controlled burns and restricting building in fire prone areas are among the strategies for reducing fire risks.

Effects on Agriculture and Ranching

Higher CO_2 concentrations and increased precipitation are likely to increase crop yields and decrease water demands, while milder winter temperatures are likely to lengthen the growing season and result in a northward shift in cropping areas.

There is the possibility that higher temperatures will also negatively affect crops by increasing heat stress, weeds, pests, and pathogens. There is a possibility that increased flooding will reduce crop production.

Fruit and nut crops, which come from perennial plants, are 32% of the value of the West's crop production, with a third of that from grapes. Since fruit and nut plants can take decades to get established, relocating such crops as an adaptive response to climate change is very likely to be more difficult than relocating annual crops.

In the ranching industry, there is a possibility that higher temperatures and increased precipitation will increase forage production and lengthen the growing and grazing season. There is also a possibility that flooding and increased incidence of animal disease will adversely affect ranching.

Adaptations: Increasing crop diversity can improve the likelihood that some crops will fare well under variable conditions, while switching to less water-demanding crops and improving irrigation efficiency would conserve water. Improved weather forecasting could aid farmers in selecting crops, timing planting and harvesting, and increasing irrigation efficiency, and aid ranchers in timing cattle sales and breeding, and in improving range management.

Shifts in Tourism and Recreation

Tourism, a growing component of the Western economy, is strongly oriented to the outdoors and sensitive to climate. Higher temperatures are likely to mean a longer season for summer activities such as backpacking, but a shorter season for winter activities, such as skiing. Ski areas at low elevations will be at risk from a shortening of the snow season and rising snowlines. There is a possibility that increases in precipitation will provide more water for sports, but some chance that there will be less water available for summer recreation and that recreation days will be limited by heat. Changes in the distribution and abundance of vegetation, fish, and wildlife will also affect recreation.

Adaptations: Strategies for tourism and recreation involve diversification of income sources. The larger, better-capitalized resorts have adapted their facilities to support winter and summer activities. These options, however, might not be available to smaller, less well-capitalized resorts.

Canadian Model

Hadley Model

- Tundra
- Taiga / Tundra
- Conifer Forest
- Northeast Mixed Forest
- Temperate Deciduous Forest
- Southeast Mixed Forest
- Tropical Broadleaf Forest
- Savanna / Woodland
- Shrub / Woodland
- Grassland
- Arid Lands

The maps are projected by the Canadian and Hadley climate models with the vegetation simulations done using a computer model called MAPPS, for the years 2070-2099.

PACIFIC NORTHWEST

The Pacific Northwest encompasses extensive forests, topography that creates abrupt changes in climate and ecosystems over short distances, and mountain and marine environments in close proximity. The Cascade Mountains divide the region climatically, ecologically, economically, and culturally. Three quarters of the region's population live west of the Cascades, concentrated in the metropolitan areas of Seattle and Portland, where the aerospace and computer industries have largely supplanted the traditional resource sectors of forestry, fishing, and agriculture. The Northwest provides a quarter of the nation's softwood lumber and plywood. The fertile lowlands of eastern Washington produce 60% of the nation's apples and large fractions of its other tree fruit.

The region has seen several decades of population and economic growth nearly twice the national rate, with population nearly doubling since 1970. The region's moderate climate, quality of life, and outdoor recreational opportunities contribute to its continuing attraction to newcomers. The same environmental attractions that draw people to the region are increasingly stressed by rapid development. Stresses arise from dam operation, forestry, and land-use conversion from natural ecosystems to metropolitan areas, intensively managed forests, agriculture, and grazing. The consequences include loss of old-growth forests, wetlands, and native grasslands; urban air pollution; extreme reduction of salmon runs; and increasing numbers of threatened and endangered species.

Observed Climate Trends

Over the 20th century, the region has grown warmer and wetter. Annual-average temperature has increased 1 to 3°F (0.5-1.5°C) over most of the region, with nearly equal warming in summer and winter. Annual precipitation has also increased across the region, by 10% on average, with increases reaching 30 to 40% in eastern Washington and Northern Idaho. The region's climate also shows significant recurrent patterns of year-to-year variability. Warm years tend to be relatively dry with low streamflow and light snowpack, while cool ones tend to be relatively wet with high streamflow and heavy snowpack. Though the differences in temperature and precipitation are small, they have clearly discernible effects on important regional resources. Warmer drier years tend to have summer water shortages, less abundant salmon, and increased probability of forest fires. These variations in the region's climate show clear correlations with two large-scale patterns of climate variation over the Pacific: the El Niño/Southern Oscillation (ENSO) on scales of a few years; and the more recently discovered Pacific Decadal Oscillation (PDO) on

Learning from Water Shortages

Seattle Public Utilities (SPU) experienced summer droughts and potential shortages in 1987, 1992, and 1998. Their responses to the three events illustrate institutional flexibility and learning. Summer 1987 began with full reservoirs, but a hot dry summer and a late return of autumn rains created a serious shortage in which water quality declined, inadequate flows were maintained

for fish, and the main reservoir fell so low that an emergency pumping station had to be installed. In response, the City developed a plan with four levels of response to anticipated shortage: advising the public of potential shortages and monitoring use; requesting voluntary use reductions; mandatory prohibitions of certain uses (such as watering lawns and washing cars); and rationing. Another drought came in 1992, following a winter with low snowpack but in which SPU had followed standard

flood-control rules by spilling water from their reservoirs. With a small snowmelt, reservoirs were low by the spring, and SPU invoked mandatory restrictions during the hot dry summer that followed. Water quality declined sharply, prompting a decision to begin building a costly ozone-purification plant.

The ill-advised spilling of early 1992 alerted SPU to the danger of following rigid reservoir rule curves, and they have since taken scales of a few decades. The observed effects of these patterns provide powerful illustrations of regional sensitivities to climate, but how they might interact with future climate change is not yet understood.

Scenarios of Future Climate

M odel scenarios project regional warming in the 21st century to be much greater than observed during the 20th century, with average warming over the region of about 3°F (1.5°C) by the 2030s and 5°F (3°C) by the 2050s. By the 2090s, average summer temperatures are projected to rise by 7-8°F (4-4.5°C), while winter temperatures rise by 8-11°F (4.5-6°C). Through 2050, average precipitation is projected to increase, although some locations have small decreases. Precipitation increases would be concentrated in winter, with little change or a decrease in summer. Because of this seasonal pattern of wetter winters and drier summers, even the projections that show annual precipitation increasing, show water availability decreasing, especially in the Hadley model. By the 2090s, projected annual average precipitation increases range from a few percent to 20% in the Hadley model, and from 20 to 50% in the Canadian model.

By the 2090s, average summer temperatures are projected to rise by 7-8°F (4-4.5°C), while winter temperatures rise by 8-11°F (4.5-6°C). Projected annual average precipitation increases range from a few percent to 20% in the Hadley model, and from 20 to 50% in the Canadian model.

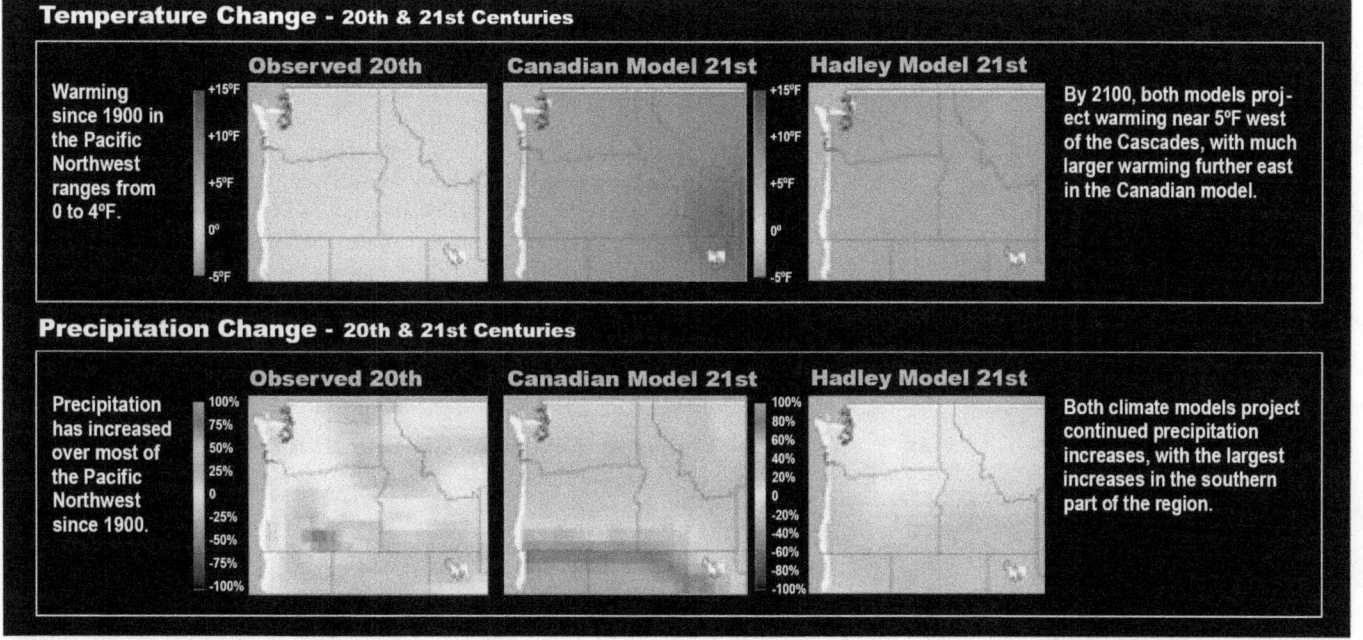

Temperature Change - 20th & 21st Centuries

Warming since 1900 in the Pacific Northwest ranges from 0 to 4°F.

Observed 20th | Canadian Model 21st | Hadley Model 21st

By 2100, both models project warming near 5°F west of the Cascades, with much larger warming further east in the Canadian model.

Precipitation Change - 20th & 21st Centuries

Precipitation has increased over most of the Pacific Northwest since 1900.

Observed 20th | Canadian Model 21st | Hadley Model 21st

Both climate models project continued precipitation increases, with the largest increases in the southern part of the region.

a more flexible approach, projecting annual supply and demand using a model including probabilistic predictions based on ENSO and PDO. During the strong El Niño of 1997-1998, SPU took early conservation education measures and allowed higher than normal reservoir fill. When 1998 brought a small snowmelt and a hot dry summer, these measures allowed the drought to pass with the public experiencing no shortage. In integrating seasonal forecasts into its operations, SPU is an uncom-monly adaptable resource-management agency. But it still has a long way to go in adapting to longer-term climate variability and change. SPU presently projects that new conservation measures will keep demand at or below present levels until at least 2010, while conservation measures and planned system expansion (including a connection with a neighboring system) will maintain adequate supply until at least 2030. Over this period, climate change is likely to have significant effects on both supply and demand, but is not yet included in planning. The warmer drier summers projected under climate change are likely to stress both supply and demand, requiring earlier capacity expansion, and triggering the more restrictive conservation measures more often. Moreover, the recent shift to cool PDO phase that has been suggested could well mask this effect for a couple of decades, risking sudden appearance of shortages when PDO next shifts back to its warm phase.

PACIFIC NORTHWEST KEY ISSUES

Changes in Timing of Freshwater Resources

Despite its reputation as a wet place, most of the Northwest receives less than 20 inches (0.5 meter) of precipitation a year, and dry summers make freshwater a limiting resource for many ecosystems and human activities. Water resources are already stressed by multiple growing demands. The projected warmer wetter winters will likely increase flooding in rainfed rivers, because there is more precipitation, and because more of it falls as rain. Projected year-round warming and drier summers will likely increase summer water shortages in both rainfed and snowfed rivers, including the Columbia, because there would be less snowpack and because it would melt earlier. In the Columbia, allocation conflicts are already acute, and the system is vulnerable to shortages.

Adaptations: Adapting to projected increases in summer shortages will likely require a combination of reducing demand, increasing supply, and reforming institutions to increase flexibility and regional problem-solving capacity. In the Colombia Basin, current infrastructure and institutions are inflexible and inadequate to deal with the projected scarcity.

Added Stresses on Salmon

While non-climatic stresses on Northwest salmon presently overwhelm climatic ones, salmon abundances have shown a clear correlation with 20th century variations in climate from decade to decade. Climate models cannot yet project the most important oceanic conditions for salmon, but the likely effects on their freshwater habitat all appear unfavorable. Increased winter flooding, reduced summer and fall flows, and rising stream and estuary temperatures are all harmful for salmon. In addition, it is possible that earlier snowmelt and peak spring streamflow will deliver juveniles to the ocean before

Increased winter flooding, reduced summer and fall flows, and rising stream and estuary temperatures are all harmful for salmon. In addition, it is possible that earlier snowmelt and peak spring streamflow will deliver juveniles to the ocean before there is adequate food for them.

Columbia Streamflow Changes

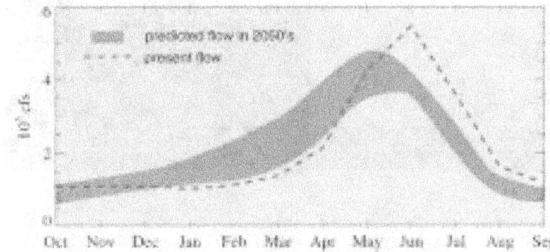

Relative to present flows (dashed), the wetter winters and drier summers simulated by climate models are very likely to shift peak streamflow earlier in the year, increasing the risk of late-summer shortages. Though the Columbia system is only moderately sensitive to climate change, allocation conflicts and a cumbersome network of interlocking authorities restrict its ability to adapt, producing substantial vulnerability to these shortages.

Observed Effects of Climate Variability on Salmon

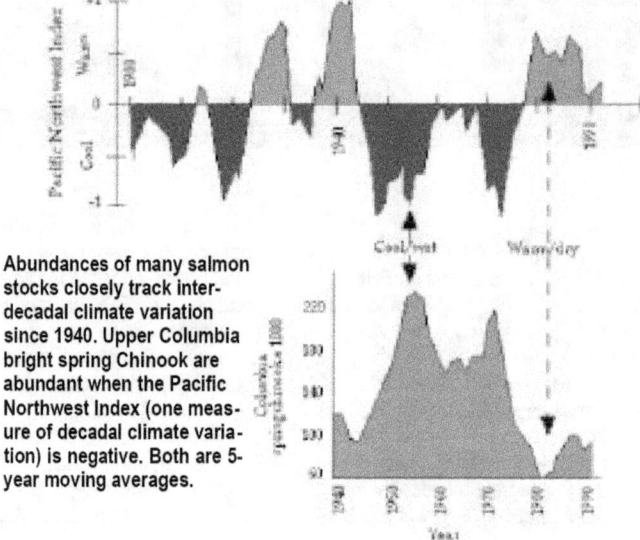

Abundances of many salmon stocks closely track inter-decadal climate variation since 1940. Upper Columbia bright spring Chinook are abundant when the Pacific Northwest Index (one measure of decadal climate variation) is negative. Both are 5-year moving averages.

there is adequate food for them. Climate change is consequently very likely to hamper efforts to restore already depleted salmon stocks,and to stress presently healthy stocks.

Adaptations: It is possible that operational changes on managed rivers would reduce current stream warming and slow future warming,although such measures will very likely be overwhelmed by continued climate warming. Measures to reduce general stress on fish,such as changing dam operations to provide adequate late-summer streamflows,might possibly increase salmon's resilience to other stresses,including climate. It is very likely that maintaining such flows will become increasingly difficult,however, under the projected regional warming that will very likely shift peak streamflows to earlier in the year. Other options include maintaining the diversity of salmon by increasing preservation of their habitat,or removing existing dams and accepting reduced ability to manage summer shortages.

Water resources are already stressed by multiple growing demands. The projected year-round warming and drier summers will likely increase summer water shortages, because there is less snowpack and because it melts earlier.

Regional Impacts:
Climate Change projected for 2050 vs observed 20th century variability

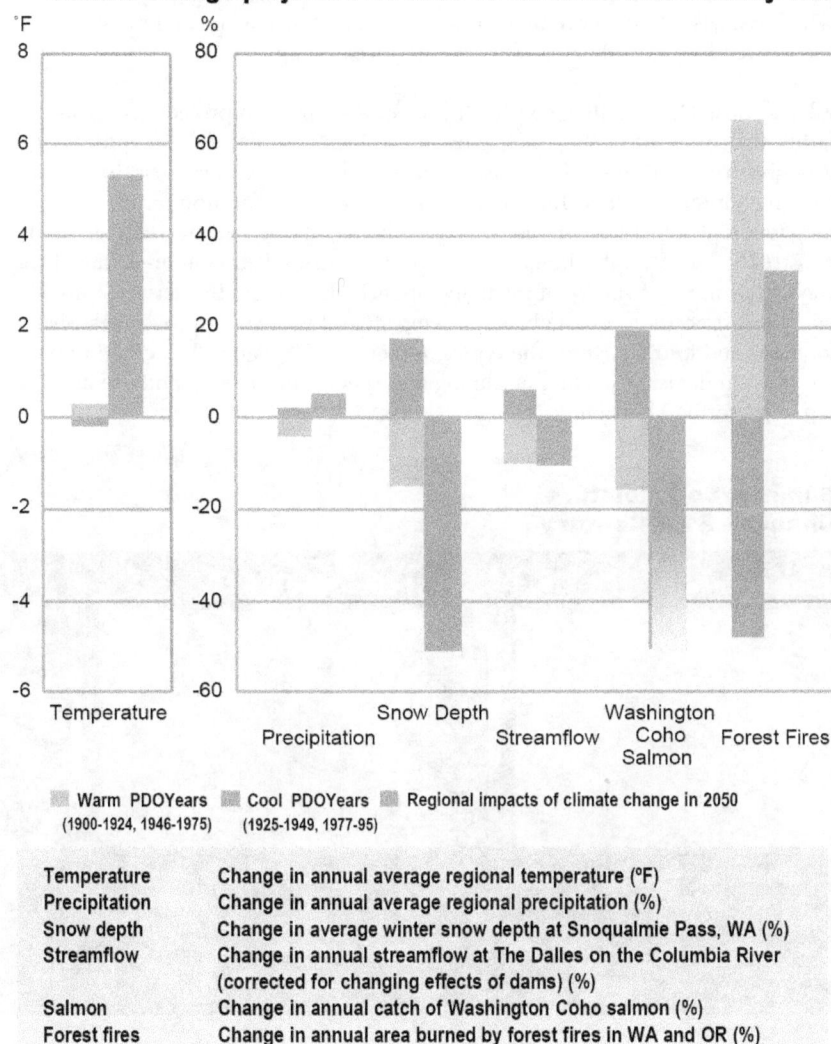

Warm PDOYears (1900-1924, 1946-1975) Cool PDOYears (1925-1949, 1977-95) Regional impacts of climate change in 2050

Temperature	Change in annual average regional temperature (°F)
Precipitation	Change in annual average regional precipitation (%)
Snow depth	Change in average winter snow depth at Snoqualmie Pass, WA (%)
Streamflow	Change in annual streamflow at The Dalles on the Columbia River (corrected for changing effects of dams) (%)
Salmon	Change in annual catch of Washington Coho salmon (%)
Forest fires	Change in annual area burned by forest fires in WA and OR (%)

This chart compares possible Northwest impacts from climate change by the 2050s with the effects of natural climate variations during the 20th century. The orange bars show the effects of the warm phase of the Pacific Decadal Oscillation (PDO), relative to average 20th century values. During warm-PDO years, the Northwest is warmer, there is less rain and snow, stream flow and salmon catch are reduced, and forest fires increase. The blue bars show the corresponding effects of cool-phase years of the PDO, during which opposite tendencies occurred.

The pink bars show projected impacts expected by the 2050s, based on the Hadley and Canadian scenarios. Projected regional warming by this time is much larger than variations experienced in the 20th century. This warming is projected to be associated with a small increase in precipitation, a sharp reduction in snowpack, a reduction in streamflow, and an increase in area burned by forest fires. Although quite uncertain, large reductions in salmon abundance ranging from 25 to 50%, are judged to be possible based on projected changes in temperature and streamflow.

PACIFIC NORTHWEST KEY ISSUES

CO$_2$ and Summer Drought Effects on Forests

Evergreen coniferous forests dominate the landscape of much of the Northwest. West of the Cascades,coniferous forests cover about 80% of the land,and include about half the world's temperate rainforest. Northwest forests have been profoundly altered by timber management and land-use conversion. These forests are quite sensitive to climate variation because warm dry summers stress them directly, by limiting seedling establishment and summer photosynthesis,as well as indirectly, by creating conditions favorable to pests and fire. The extent,species mix,and productivity of Northwest forests are likely to change under projected 21st century climate change,but the specifics of these changes are not known with confidence at present. They are very likely to depend on interactions between the timing and amount of precipitation,the seasonal water-storage capacity of forest soils,and changes in trees' water-use efficiency under elevated CO$_2$. It is very likely that these factors will jointly determine the consequences of the likely increase in summer moisture stress, which will also depend on interactions with forest management practices,land-use conversion,and other pressures from development.

Adaptations: Options include planting species adapted to projected climate rather than present climate;managing forest density to reduce susceptibility to drought stress and fire risk;and using prescribed burning to reduce the risk of large,high-intensity fires. Increased capacity for long-term monitoring and planning would likely help with management. Reduced tree cutting, reduced road construction,and establishment of large buffers around streams are some of the ways to promote diversity of plant and animal species and the services provided by forest ecosystems (such as purifying air and water). Improved seasonal forecasts,and knowledge of the typical effects of ENSO and PDO, could possibly assist in decision making on timing and species of planting,and use and timing of prescribed burning.

Northwest forests are quite sensitive to climate variation because warm dry summers stress them directly, by limiting seedling establishment and summer photosynthesis, as well as indirectly, by creating conditions favorable to pests and fire. The extent, species mix, and productivity of Northwest forests are likely to change under projected 21st century climate change...

Summer Soil Moisture Change - 21st Century

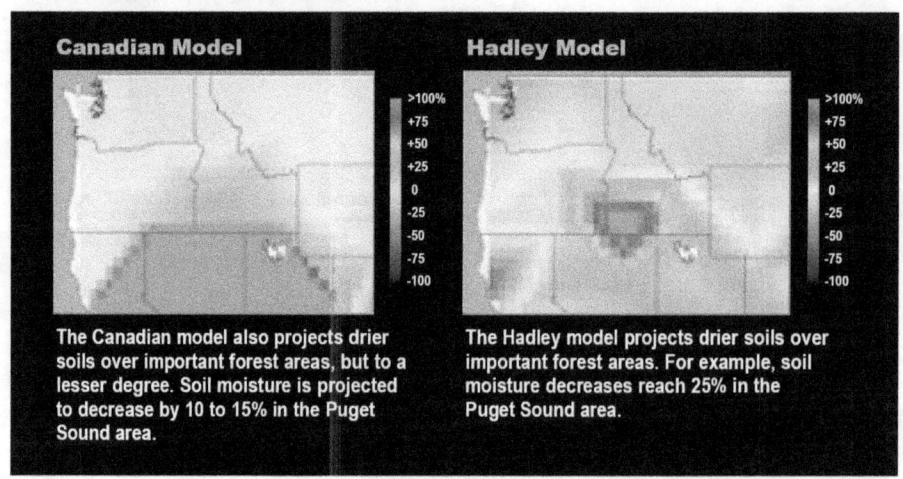

Canadian Model

The Canadian model also projects drier soils over important forest areas, but to a lesser degree. Soil moisture is projected to decrease by 10 to 15% in the Puget Sound area.

Hadley Model

The Hadley model projects drier soils over important forest areas. For example, soil moisture decreases reach 25% in the Puget Sound area.

Sea-level Rise Impacts on Coastal Erosion

Sea-level rise is likely to require substantial investments in order to avoid coastal inundation, especially in the low-lying communities of southern Puget Sound where coastal subsidence is occurring. Other likely effects include increases in winter landslides, and increased erosion on sandy stretches of the Pacific Coast. Severe storm surges and erosion are presently associated with El Niño events, which raise sea level for several months and change the direction of prevailing winds. Climate change is projected to bring similar shifts. Projected heavier winter rainfall is likely to increase soil saturation, landsliding, and winter flooding. All these changes would likely increase the danger to property and infrastructure on bluffs and beachfronts, and beside rivers.

Adaptations: The current coastal management system is not particularly adaptable, even to current climate variability and risks, and there is little inclination to restrict development in vulnerable locations. Adaptation strategies would involve conserving remaining natural coastal areas, placing less property at risk in low-lying or flood- or slide-prone areas, assigning more of the associated risk to property owners through insurance rates, and more effective transfer of climate change information to local governments, where most planning authority lies.

Severe storm surges and erosion are presently associated with El Niño events, which raise sea level for several months and change the direction of prevailing winds. Climate change is projected to bring similar shifts.

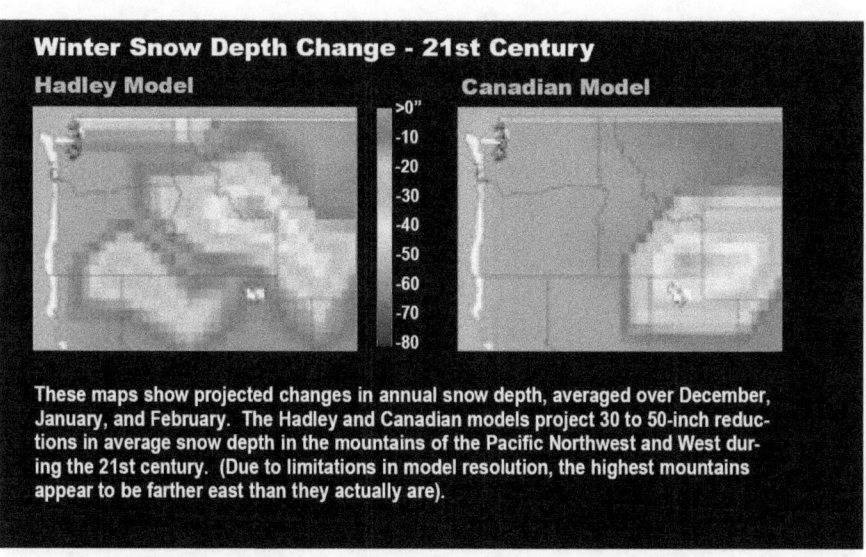

Winter Snow Depth Change - 21st Century

Hadley Model **Canadian Model**

>0"
-10
-20
-30
-40
-50
-60
-70
-80

These maps show projected changes in annual snow depth, averaged over December, January, and February. The Hadley and Canadian models project 30 to 50-inch reductions in average snow depth in the mountains of the Pacific Northwest and West during the 21st century. (Due to limitations in model resolution, the highest mountains appear to be farther east than they actually are).

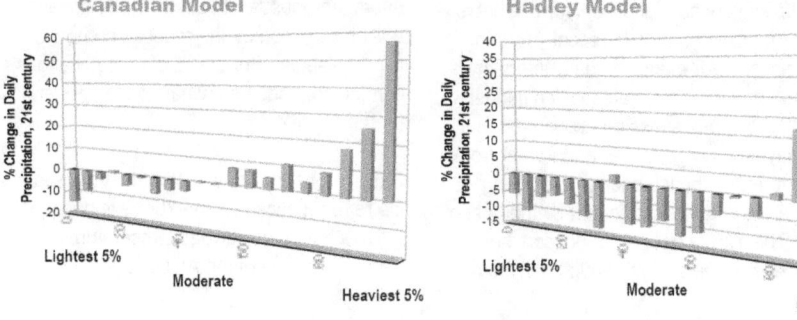

Projected Northwest Daily Precipitation Change - 21st Century

Canadian Model   Hadley Model

Lightest 5%   Lightest 5%
Moderate   Moderate
Heaviest 5%   Heaviest 5%

Days receiving precipitation, sorted by percentile.

Over the 21st century, both models project increases in annual average precipitation and in extreme precipitation events. As the graphs for both models show, the largest precipitation increases are projected to occur on days already receiving the most.

ALASKA

Alaska spans an area nearly a fifth the size of the entire lower 48 states,and includes a wide range of physical, climatic,and ecological diversity in its rainforests,mountain glaciers,boreal spruce forest,and vast tundra,peatlands, and meadows. It contains 75% by area of US national parks and 90% of wildlife refuges,63% of wetlands,and more glaciers and active volcanoes than all other states combined. Direct human pressures on the state's land environment are light,but pressures on its marine environment from large commercial fisheries are substantial. Lightly populated (614,000 people) and growing about 1.5% per year,Alaska has the nations' highest median household income,with an economy dominated by government (44% of incomes) and natural resources (oil 35%, fisheries 7%). Diverse subsistence livelihoods,practiced primarily by native communities,depend on fish,marine mammals,and other wildlife,and play a social and cultural role vastly greater than their contribution to monetary incomes.

Observed Climate Trends

Alaska has warmed substantially over the 20th century, particularly over the past few decades. Average warming since the 1950s has been 4°F (2°C). The largest warming, about 7°F (4°C),has occurred in the interior in winter. The growing season has lengthened by more than 14 days since the 1950s. Some records suggest that much of the recent warming occurred suddenly around 1977. Alaska has also grown wetter recently, with precipitation over most of the state increasing 30% between 1968 and 1990. The observed warming is part of a larger trend through most of the Arctic corroborated by many

KEY ISSUES

- Permafrost Thawing and Sea Ice Melting

- Increased Risk of Fire and Insect Damage to Forests

- Sensitivity of Fisheries and Marine Ecosystems

- Increased Stresses on Subsistence Livelihoods

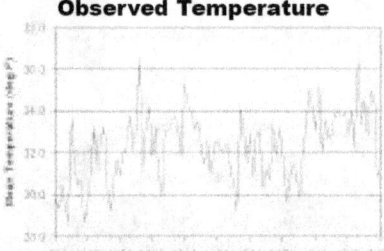
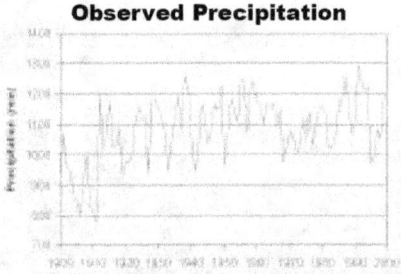

Over the 20th century, average temperature and precipitation in Alaska have both increased.

"Everything is Tied to Everything Else" A Lesson from Alaska

Caleb Pungowiyi is a Yupik Eskimo who lives in the Arctic, moving back and forth from Alaska to Siberia in pursuit of walrus and other sea mammals. Gathering food directly from the land and the sea makes

the Yupiks very careful observers of what is going on around them. In recent years they have noticed, for example, that winters are warmer, that the walrus are looking thinner and their blubber is less nutritious, and that they have had to go further and further from shore to reach the ice pack where young seals are being fed fish caught by their parents. The Yupiks have even noticed that some killer whales have begun eating sea

otters, an unusual shift in their diet apparently brought on by the reduced number of fish and seals. But are all of these changes connected, and, if so, what do they portend for the future?

Satellite observations confirm that the sea ice retreat noticed by the Yupiks is happening much more widely, as temperatures warm over most of the Arctic region.

independent measurements of sea ice, glaciers, permafrost, vegetation, and snow cover. In contrast to other regions, the most severe environmental stresses in Alaska at present are climate-related.

Scenarios of Future Climate

Models project that rapid Arctic warming will continue. For Alaska, the Hadley and Canadian models project 1.5-5°F (1-3°C) more warming by 2030, and 5-12°F (3-6.5°C) (Hadley) or 7-18°F (4-10°C) (Canadian) by 2100. The warming is projected to be strongest in the north and in winter. Both models also project continued precipitation increases in most of the state reaching 20-25% in the north and northwest, with areas of up to 10% decrease along the south coast. Projections indicate that increased evaporation from warming will more than offset increased precipitation, however, making soils drier throughout most of the state.

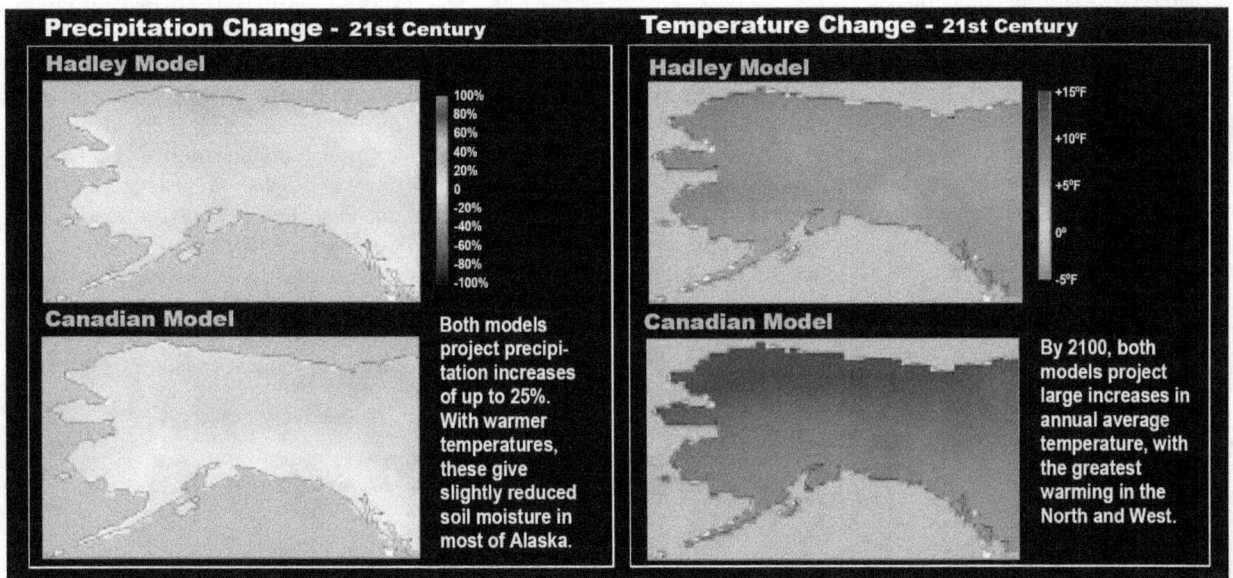

Precipitation Change - 21st Century

Hadley Model

100%
80%
60%
40%
20%
0
-20%
-40%
-60%
-80%
-100%

Canadian Model

Both models project precipitation increases of up to 25%. With warmer temperatures, these give slightly reduced soil moisture in most of Alaska.

Temperature Change - 21st Century

Hadley Model

+15°F
+10°F
+5°F
0°
-5°F

Canadian Model

By 2100, both models project large increases in annual average temperature, with the greatest warming in the North and West.

Because the edge of the sea ice is further out to sea in deeper water, walrus – which rest on the ice and feed on the bottom – must dive deeper to feed and find less food, causing their weakened condition. Because sea ice is melting back earlier in the year, the seal pups being raised on the edge are smaller when they must leave the ice, worsening their chance of survival. With fewer seal pups, sea otters become an alternative

food source for whales. Because a favorite food of sea otters is sea urchins, fewer sea otters will mean more sea urchins. Sea urchins' favorite food is the kelp that provide the breeding grounds for the fish, so more sea urchins will mean less kelp and thus fewer fish. And with walrus and seal populations declining, it is these very fish that the Yupik need more than ever to feed themselves.

It may seem like only a little warming in a very cold place, but for the Yupiks, the warming is significantly disrupting their traditional food sources because as Caleb Pungowiyi says, in their environment, like all environments, "everything is tied to everything else."

75

ALASKA KEY ISSUES

Permafrost Thawing and Sea Ice Melting

The rapid warming Alaska is already experiencing is bringing substantial ecological and socioeconomic impacts, many of which result from thawing permafrost or melting sea ice. Permafrost underlies most of Alaska, and the recent several decades of warming have been accompanied by extensive thawing, causing increased erosion, landslides, sinking of the ground surface, and disruption and damage to forests, buildings, and infrastructure. Thawing is projected to accelerate under future warming, with as much as the top 30 to 35 feet (10 meters) of discontinuous permafrost thawing by 2100. Warming is also likely to impair transport by shortening the seasonal use of ice roads.

Sea ice off the Alaskan coast is retreating and thinning, with widespread effects on marine ecosystems, coastal climate, human settlements, and subsistence activities. The area of multi-year Arctic sea ice has decreased 14% since 1978, with an apparent sharp increase in the annual rate of loss in the 1990s. Since the 1960s, sea ice over large areas of the Arctic basin has thinned by 3 to 6 feet (1 to 2 meters), losing about 40% of its total thickness. All climate models project large continued loss of sea ice, with year-round ice disappearing completely in the Canadian model by 2100.

Retreat of sea ice allows larger storm surges to develop, increasing the risk of inundation and increasing erosion on coasts that are also made vulnerable by permafrost thawing. In some regions, shorelines have retreated more than 1500 feet (400 meters) due to erosion, over the past few decades. Several Alaskan coastal villages will soon have to be fortified or relocated. Loss of sea ice also causes large-scale changes in marine ecosystems, threatening populations of

marine mammals and polar bears that depend on ice, and the subsistence livelihoods that depend on them. It is possible that further retreat of sea ice will also bring some benefits, principally to ocean shipping and offshore oil exploration and extraction, and will have major implications for trade and national defense.

Adaptations: Adaptations to thawing depend on the site. Minimizing surface disruption and heat transfer from buildings can reduce local contributions to thawing. Selecting sites without ice-rich permafrost, where feasible, can reduce the likelihood of subsidence. Otherwise, structures' vulnerability to thawing can only be reduced through such costly measures as building on very deep or refrigerated piles, or alternatively by stripping surface soil five years or more in advance, in order to let thawing occur before construction. Coastal settlements can be fortified or moved inland, but these options are likely to be expensive. No effective protection is likely to be available for forests or natural coastlines.

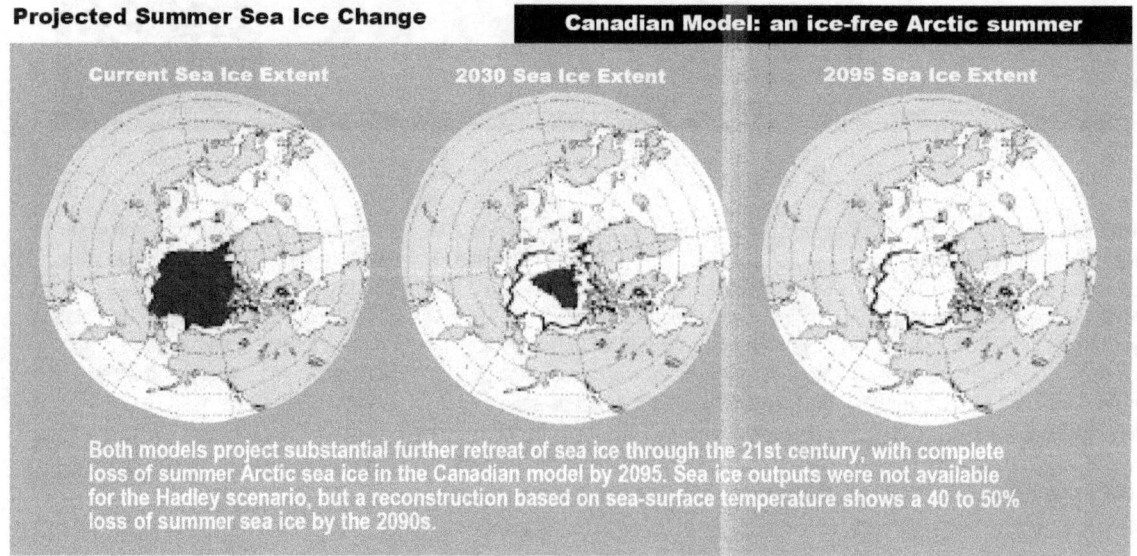

Projected Summer Sea Ice Change

Canadian Model: an ice-free Arctic summer

Current Sea Ice Extent   2030 Sea Ice Extent   2095 Sea Ice Extent

Both models project substantial further retreat of sea ice through the 21st century, with complete loss of summer Arctic sea ice in the Canadian model by 2095. Sea ice outputs were not available for the Hadley scenario, but a reconstruction based on sea-surface temperature shows a 40 to 50% loss of summer sea ice by the 2090s.

Increased Risk of Fire and Insect Damage to Forests

One third of Alaska is covered by forests that support subsistence livelihoods, recreation, and in the southeastern part of the state, a timber industry. Recent warming has increased average growing degree-days by about 20% over the state, apparently increasing productivity where forests are not moisture limited (mainly on the southern coast), but reducing productivity where they are (in much of the interior). The climate increasingly favors expansion of boreal forest into the tundra zone, particularly on the Seward Peninsula. Recent warming has also been accompanied by unprecedented increases in forest disturbances, including insects, blow-downs and fire. A sustained infestation of spruce bark beetles, which in the past have been limited by cold, has caused widespread tree deaths over 2.3 million acres on the Kenai Peninsula since 1992, the largest loss to insects ever recorded in North America. At the same time, increases in blow-downs from intense windstorms, and in canopy breakage from the heavy snows typical of warm winters may have increased vulnerability of forests to insect attack. Significant increases in fire frequency and intensity, both related to summer warming, have also occurred. Simultaneously, the potential damage from forest fires has increased due to a rapid increase in dispersed human settlement in forests. The projected further warming is likely to increase risk of both fire and insect disturbances, even in the near term. In the longer term, large-scale transformation of landscapes is possible, including expansion of boreal forest into the tundra zone, shifts of forest types due to fire and moisture stress, northward expansion of some commercially valuable species, and the appearance of significant fire risk in the coastal forest for the first time since observations began. In present commercial forests, management practices must adapt to heightened fire and pest risk, including potential interactions between them. In the longer term, there is some chance that northward shift of forest productivity and commercially valuable species will hold substantial opportunities for new commercial timber development.

In the longer term, large-scale transformation of landscapes is possible, including expansion of boreal forest into the tundra zone, shifts of forest types due to fire and moisture stress, northward expansion of some commercially valuable species, and the appearance of significant fire risk in the coastal forest for the first time since observations began.

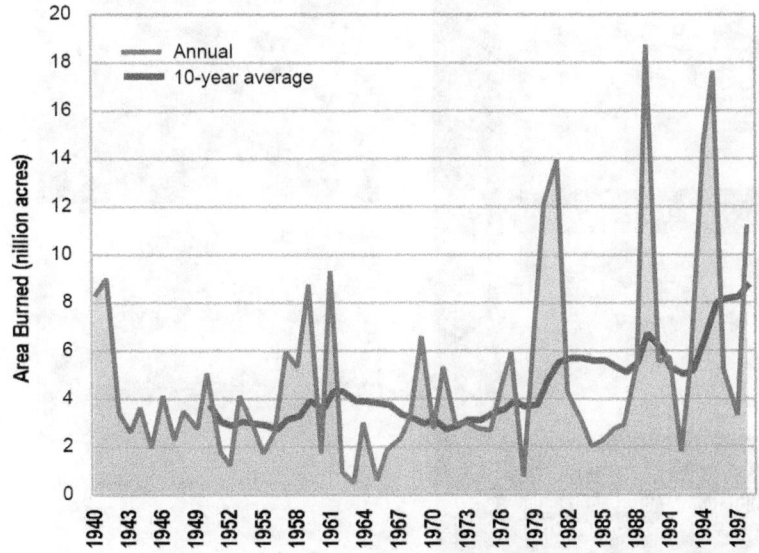

Annual Area of Northern Boreal Forest Burned in North America

Legend:
— Annual
— 10-year average

Y-axis: Area Burned (million acres)
X-axis: 1940 – 1997

The Alaskan boreal forest is a small part of an enormous forest that extends continuously across the northern part of North America. The average area of this forest burned annually has more than doubled since 1970.

Alaska's fisheries, the largest in the nation, appear to have shown substantial sensitivity to climate fluctuations over the 20th century.

ALASKA KEY ISSUES

Sensitivity of Marine Ecosystems and Fisheries

The Gulf of Alaska and Bering Sea support marine ecosystems of great diversity and productivity, and the nation's largest commercial fishery. In 1995, Alaskan fisheries landed 2.1 million tons ($1.45 billion worth), representing 54% of the landings and 37% of the value of all US fisheries. The productivity of these ecosystems fluctuates with year-to-year and especially decade-to-decade climate variability. Some data suggest that climate fluctuations have caused extreme regime shifts in these ecosystems several times since 1900, most recently in the late 1970s and per-

haps again in the 1990s. Salmon stocks soared in 1977 and most groundfish stocks a few years later, while forage fish such as capelin and herring declined sharply, bringing subsequent declines in the seabirds and marine mammals that feed on them. These changes likely reflect joint effects of climate fluctuations, ocean circulation, and human harvesting. Consequently, while the effect of projected climate change on these ecosystems is likely to be large, little is known of its specific character.

Adaptations: Potential adaptation measures include reducing the specialization of the fisheries capital equipment to particular species in particular places, to increase the industry's robustness to potential shifts in species' location and abundance; increasing the flexibility of fishing regulations through such

measures as variable quotas or buybacks; and limiting other ecosystem stresses such as marine pollution or disruption of nursery grounds.

Increased Stress on Subsistence Livelihoods

Subsistence makes an important contribution to livelihoods in many isolated rural communities, especially but not exclusively for native peoples. Subsistence is practiced to gather food, but is also important to health, culture, and identity. Alaska's 117,000 rural residents collect about 43 million pounds of wild food annually, equivalent to 375 pounds per person each year. In some remote communities, the subsistence harvest is as high as 800 pounds per person. Fish comprise 60% of the wild harvest, but there is substantial regional variation: west

Summer Soil Moisture Change

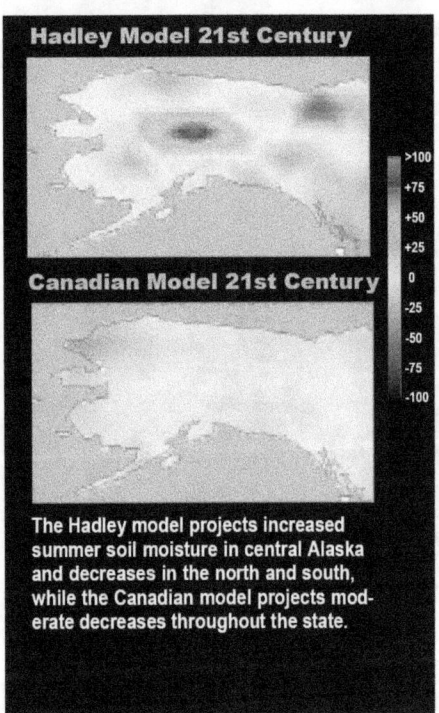

The Hadley model projects increased summer soil moisture in central Alaska and decreases in the north and south, while the Canadian model projects moderate decreases throughout the state.

Winter Maximum Temperature Change

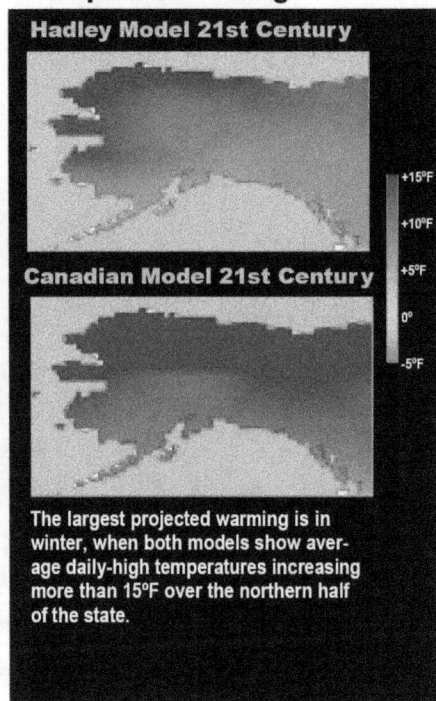

The largest projected warming is in winter, when both models show average daily-high temperatures increasing more than 15°F over the northern half of the state.

Habitats for wildlife, on land and in the ocean, will come under increasing pressure from vegetation shifts and sea ice retreat.

coast communities rely principally on fish, while northern ones rely more on marine mammals, and interior ones on both fish and land mammals. Present climate change already poses serious harms to subsistence livelihoods. Many populations of marine mammals, fish, and seabirds have been reduced or displaced. Reduced snow cover, a shorter river ice season, and thawing of permafrost all obstruct travel to harvest wild food. Retreat and thinning of sea ice, with associated stress on marine mammal and polar bear populations and increased open-water roughness, have made hunting more difficult, more dangerous, and less productive. It is possible that projected near-term climate changes will enhance certain subsistence harvests, but in general are likely to intensify present harms, through further loss of sea ice, river ice, and permafrost. In the longer term, projected ecosystem shifts are likely to displace or change the resources available for subsistence, requiring communities to change their practices or move. Shifts in the composition of tundra vegetation may decrease nutrition available for caribou and reindeer, while invasion of the tundra by boreal or mixed forest is likely to curtail the range of caribou and musk-ox.

Adaptations: Although subsistence cultures have historically adapted to climate variability by shifting practices and target species, subsistence practices are now both hotly contested and extensively regulated, posing challenges to traditional means of adaptation. It is possible that projected climate change will overwhelm the available responses, particularly for communities that rely on marine mammals. Some communities may be forced to reduce their dependence on the wild harvest, or relocate. General measures to increase incomes may mitigate some impacts, on nutrition for example, but not the cultural effects of lost subsistence resources.

Simulated Vegetation Distribution

Current

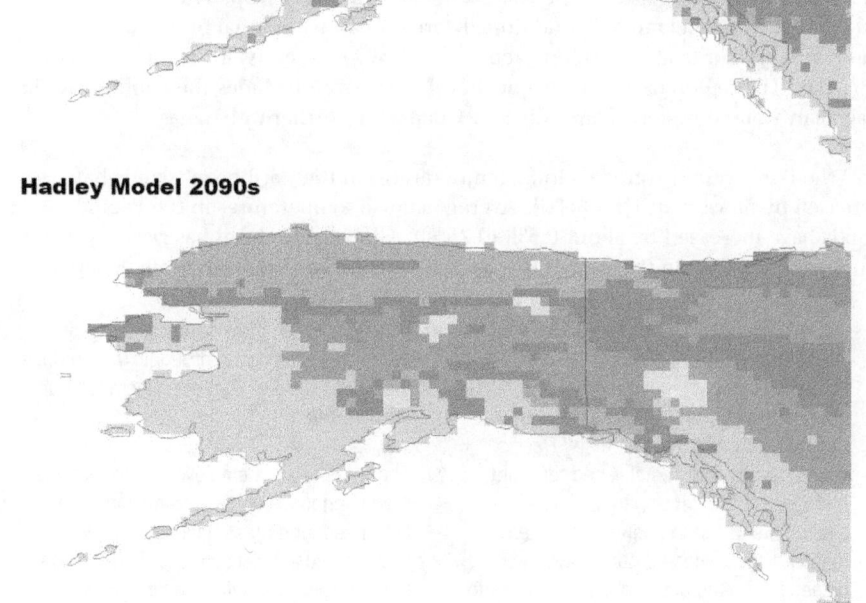

Tundra
Taiga / Tundra
Boreal Conifer Forest
Temperate Evergreen Forest
Temperate Mixed Forest
Tropical Broadleaf Forest
Savanna / Woodland
Shrub / Woodland
Grassland
Arid Lands

Hadley Model 2090s

Under the Hadley scenario, the MAPSS biogeography model projects large-scale loss of tundra and taiga ecosystems as forests expand north and west. Likely consequences include disruption of wildlife migration and associated subsistence livelihoods, as well as the potential for large releases of soil carbon.

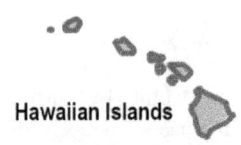

Hawaiian Islands

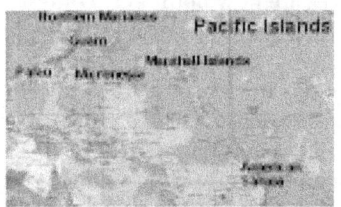

Pacific Islands

Northern Marianas
Guam
Palau   Micronesia
Marshall Islands
American Samoa

U.S. Virgin Islands

The scope of this section includes the US-affiliated islands of the Caribbean and Pacific. In the Caribbean, this includes Puerto Rico and the US Virgin Islands. In the Pacific, it includes the Hawaiian Islands, American Samoa, the Commonwealth of the Northern Mariana Islands, Guam, the Federated States of Micronesia, the Republic of the Marshall Islands, and the Republic of Palau.

KEY ISSUES

- Freshwater Resources

- Public Health and Safety

- Ecosystems and Biodiversity

- Sea-level Variability

ISLANDS IN THE CARIBBEAN AND THE PACIFIC

The Caribbean and Pacific islands that are affiliated with the US provide a unique setting for consideration of climate variability and change. Islands contain diverse and productive ecosystems,and include many specialized and unique species. After centuries of depending on subsistence agriculture and fishing,island economies are now based heavily on tourism,tuna processing and transshipment, and agricultural production for export (including sugar cane,bananas,pineapple, spices,and citrus fruits),making them highly responsive to external economic forces. The stability of these economies is also dependent on the health of the unique natural resources,all of which are sensitive to climate.

Many islands are facing the stresses of rapid human population growth,increasing vulnerability to natural disasters,and degradation of natural resources.Droughts and floods are among the climate extremes of most concern as they affect the amount and quality of water supplies in island communities and thus can have significant health consequences.Due to their small size and isolation,many islands face chronic water shortages and problems with waste disposal. Some are facing a species extinction crisis; for example,the Hawaiian Islands have the highest extinction rate of any state in the nation. For most island communities,infrastructure and economic activities are located near the coast,making them highly vulnerable to storm events and sea-level fluctuations.

Observed Climate Trends

In the Pacific,tropical storms and typhoons are common between May and December, but can occur in any month. In the Caribbean,the hurricane season spans the months from June to November. The El Niño Southern Oscillation (ENSO) cycle affects sea level, rainfall,and cyclone activity (hurricanes or typhoons, depending on the region). In the Caribbean, Atlantic hurricanes are suppressed during El Niño,while they increase during La Niña. In the Pacific,during El Niño events,Hawaii,Micronesia,and the islands of the southwest tropical Pacific often receive below normal rainfall. Additionally, areas of above normal precipitation, along with greater tropical cyclone activity, typically shift eastward towards French Polynesia. The region of greater tropical cyclone activity includes the central Pacific (Hawaiian waters),eastern Marshalls and Guam,and northern Marianas.

Over the last century, average annual temperatures in the Caribbean islands have increased by more than 1°F (0.5°C). Average annual temperatures in the Pacific Islands have increased by about 0.5°F (0.25°C). Globally, sea level has risen by 4 to 8 inches (10-20 cm) in the past 100 years with significant local variation. Relative

The Value of Climate Forecasts

The 1997-1998 El Niño event offers a vivid example of how information about potential consequences can be used to support decision making and benefit society. In 1997, the Pacific ENSO Applications Center (based in Hawaii and Guam) provided early forecasts of El Niño-related droughts in

Hawaii, Micronesia, and the tropical southwest Pacific. The Applications Center subsequently pursued an aggressive program of government briefings, public education, and outreach. As a result, many Pacific Island governments established "drought task forces" and developed mitigation plans. In addition to addressing governmental actions, these drought task forces helped inform the public about strategies to

conserve water, prevent outbreaks of diseases associated with droughts, and reduce the risk of wildfires that often increase during droughts.

These task forces employed radio and television announcements, information hotlines, brochures, and presentations on El Niño and drought in local schools. In response to the public information campaign, water sea level,which also takes into account natural and human-caused changes in the land elevation such as tectonic uplifting and land subsidence (sinking),is also showing an upward trend (3.9 inches, about 10 cm,per 100 years) at sites monitored in the Caribbean and Gulf of Mexico. Although absolute sea level is also rising in the Pacific,trends vary greatly from island to island due to the fact that some islands are rising;ENSO and other short-term variations further complicate the picture. Low-lying islands that are not rising are very likely to be at risk from sea-level rise.

Some models suggest more persistent El Niño-like conditions across the Pacific. This would lead to a reduction of fresh water resources.

Scenarios of Future Climate

Pacific and Caribbean islands will possibly be affected by: changes in patterns of natural climate variability (such as ENSO); changes in the frequency, intensity, and tracks of tropical cyclones;and changes in ocean currents. These islands are very likely to experience increasing air and ocean temperatures and changes in sea level (including storm surges and sustained rise). Some recent climate model studies also project that ENSO extremes are likely to increase with increasing greenhouse gas concentrations. Some models suggest more persistent El Niño-like conditions across the Pacific. This would lead to a reduction of fresh water resources in areas of the western Pacific, Micronesia,and the southwest tropical Pacific,and a reduction in Atlantic hurricane frequency.

One hurricane modeling study suggests that peak wind speed will increase by 5-10% by the end of the 21st century along with significant increases in peak precipitation rates. Apart from the linkage with ENSO, there is significant uncertainty about how increasing global temperatures will affect hurricane and typhoon frequency and tracks.

Islands are home to unique ecosystems and species, with unsurpassed biodiversity. These resources are already threatened by invasive non-native plant and animal species, as well as urban expansion and various industrial activities, resulting in the highest extinction rates of all regions of the US. They can also be highly climate-sensitive.

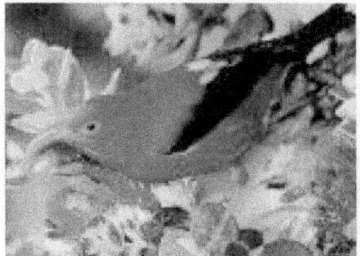
Endangered Hawaiian hakalau.

Coral reef ecosystem.

Mangrove.

management agencies implemented water conservation plans and repaired their systems; residents repaired their water catchment systems; and local vendors supplied new catchment systems and new water storage tanks.

Even with these precautionary measures, the 1997-1998 El Niño produced such extensive drought conditions that water rationing became necessary, limited hours of water use were imposed on most islands, and eventually water augmentation was required on several Pacific islands.

Agriculture suffered from the droughts everywhere except on Guam, where there was ample water for irrigation.

Still, the consequences could have been worse. Advance warning through emerging forecasting capabilities and a focused program of education and outreach clearly helped mitigate the negative impacts. These actions prevented death and greatly reduced suffering to less than occurred during the 1983 El Niño.

ISLANDS IN THE CARIBBEAN AND THE PACIFIC KEY ISSUES

Freshwater Resources

Adequate water supplies are critical for the well-being and economic security of the islands and are needed for tourism, agriculture, fish processing,and urban/municipal users,as well as natural ecosystems. On many islands, water resources or access to them are already limited and subject to competing demands. It is possible that climate change and the resulting sea-level rise will adversely affect water supplies in the future through more frequent droughts, floods,and salt water intrusion into freshwater lenses. The populations of the Pacific Islands are primarily concerned with future conditions that are likely to exacerbate drought. In Puerto Rico and the US Virgin Islands drought is also a concern,as are flooding and landslides associated with heavy precipitation events.

Adaptations: Strategies for providing adequate water resources include improved rainfall catchment;

improved storage and distribution systems;development of under-utilized or alternative sources,better management of water supply and infrastructure;increased water conservation programs;construction of groundwater recharge basins for runoff;more effective use of ENSO forecast information;and application of new technology, such as desalinization. For agriculture,strategies include exploring the feasibility of planting more drought-resistant crops.Consideration of the effects of climate change and variability on freshwater resources should be integrated into community planning and tourism develpment.

Public Health and Safety

Both coastal and inland island populations and infrastructure are already at risk from climate extremes. Storms can damage or destroy buildings,damage infrastructure,and disrupt public services. Both the Pacific and Caribbean regions are familiar with severe cyclones,which have caused billions of dollars in damage from the destruction of housing, agriculture,

roads and bridges,and lost tourism revenue. In both regions,a large percentage of people,infrastructure,and economic activities are located near the coast,leading to dense areas of vulnerability. The unique topography of Puerto Rico and the Virgin Islands makes them susceptible to floods and landslides often resulting from severe storms. It is possible that the frequency of extreme events may increase over the next few decades to a century thereby increasing the risk to public health and safety.

Adaptations: Strategies include upgrading and protection of infrastructure,comprehensive disaster management programs, changes in land use policies,and adoption and enforcement of more stringent building codes. For reducing public health risks,strategies include improved sanitation and health care infrastructure, emergency plans,and public education about health risks posed by floods and droughts.

Freshwater Lens Effect in Island Hydrology

On many islands, the underground pool of freshwater that takes the shape of a lens is a critical water source. The freshwater lens is suspended by salt water. If sea level increases, and/or if the lens becomes depleted because of excess withdrawals, salt water from the sea can intrude, making the water unsuitable for many uses. The size of the lens is directly related to the size of the island: larger islands have lenses that are less vulnerable to tidal mixing and have enough storage for withdrawals. Smaller island freshwater lenses shrink during prolonged periods of low rainfall, and water quality is easily impaired by mixing with salt water. Short and light rainfall contributes little to recharge of these sources. Long periods of rainfall are needed to provide adequate recharge.

Path of Hurricane Georges in Relation to Puerto Rico with Precipitation Totals

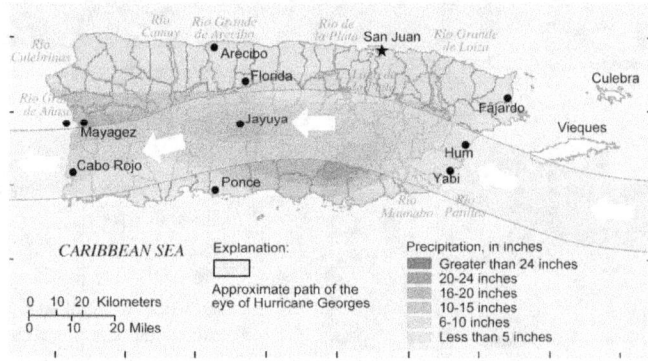

On September 21, 1998, Hurricane Georges swept across Puerto Rico. The eye of the hurricane was 25-30 miles wide and passed within 15 miles of the capital, San Juan, leaving a trail of devastation in its wake. The path of the hurricane and rainfall totals are shown here. Some areas received up to 26 inches of rain within 24 hours. Flooding, landslides, and catastrophic losses in infrastructure resulted.

Ecosystems and Biodiversity

The isolation of islands has made them living laboratories for understanding species adaptation and evolution,but has also made them extremely vulnerable to invasive species and other stresses. Islands are home to unique ecosystems and species,with unsurpassed biodiversity. These resources are already threatened by invasive non-native plant and animal species,as well as urban expansion and various industrial activities, resulting in the highest extinction rates of all regions of the US. They can also be highly climate-sensitive. For example,coral bleaching associated with El Niño events and the long-term warming of surface waters has become widespread in both the Pacific and Caribbean since the 1990s. During the El Niño of 1997-98,coral bleaching in Palau, known for its spectacular coral reefs, was extensive.

Other possible concerns include increased extinction rates of mountain species that have limited opportunities for migration;increased rates of changes in mangrove ranges and health;and declines in forests due to floods,droughts,or increased incidence of pests,pathogens,or fire. It is possible that increases in the frequency or intensity of hurricanes would generally favor invasive species. In addition,the unique "cloud forests"located on some of the islands occupy a narrow geographical and climatological niche. A slight shift in temperature or precipitation patterns would possibly cause this zone to shift upwards enough to be eliminated.

Adaptations: While options are limited,strategies include efforts at slowing biological invasions,strengthening and enforcing policies that protect critical habitats,improving understanding of the local effects of climate variability and change,and increasing the awareness of tourists and the public concerning the value of species and biodiversity.

Sea-level Variability

Sea-level rise,both long-term and episodic,is already an extremely important issue for many of the islands.Sea-level rise results in coastal erosion.inundation,and salt water intrusion into freshwater lenses and coastal agricultural zones (where taro,pulaka,and yams are grown). Future sea-level rise,both global and due to episodic events (such as extreme lunar tides,ENSO-related changes,and storm-related wave conditions) will increasingly contribute to negative consequences for island populations and ecosystems. Most at risk are low-lying islands and atolls. Examples of sites that are already close to sea level include the Republic of the Marshall Islands in the Pacific and much of the metropolitan area of San Juan in Puerto Rico.

Adaptations: Strategies include efforts to protect coastal infrastructure,transportation and water systems, agriculture,and communities; integrated coastal zone management; and crop diversification and the use of salt-resistant crops. Retreat from risk prone areas is likely to be necessary in some cases,but will be complicated due to land ownership,and could have significant consequences for social and cultural identity.

El Niños and La Niñas
Observed and Projected Sea Surface Temperature Anomalies

Historical (1860-1990)

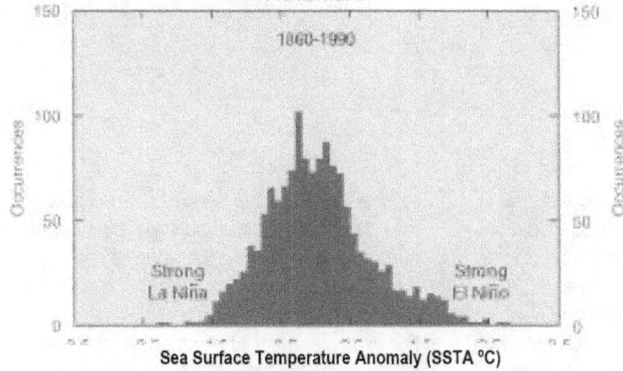

With Projected Increases of
CO₂ and Sulfates (1990-2100)

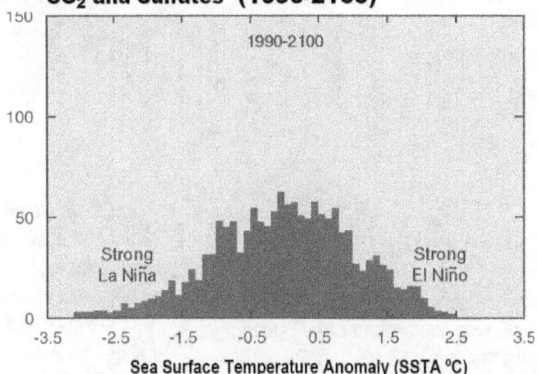

Some model projections suggest more frequent El Niño-like conditions and stronger La Niñas as a result of climate change. Sea surface temperature (SST) deviations from normal in the equatorial Pacific are used to measure the strength of El Niños and La Niñas. These high resolution model projections by the Max Planck Institute suggest more SST deviations from normal and thus more frequent El Niños and stronger La Niñas in the future. The high bars in the center are occurrences of normal SSTs. In the projections in the right hand graph, these normal temperatures occur less frequently, while lower (La Niña) and higher (El Niño) SSTs occur more frequently. The Max Planck model is used here because it has been able to reproduce the strength of these events better than other models due to its physics and ability to resolve fine scale structure in the ocean.

NATIVE PEOPLES AND HOMELANDS

American Indians and the indigenous peoples of Alaska, Hawaii, and the Pacific and Caribbean islands comprise almost 1% of the US population. The federal government recognizes the unique status of more than 565 tribal and Alaska Native governments as "domestic dependent nations." The relationships between these tribes and the federal government are determined by treaties, executive orders, tribal legislation, acts of Congress, and decisions of the federal courts. These agreements cover a range of issues that will be important in facing the prospects of climate change, from responsibilities and governance, to use and maintenance of land and water resources.

Of the approximately 1.9 million people formally enrolled in federally recognized tribes, over half live on hundreds of reservations throughout the country. Within the 48 conterminous states, tribal lands total about 56 million acres, an area about the size of the state of Minnesota. Those who do not live on tribal lands, but instead live in cities, suburbs, and small rural communities across the US, will face the same set of challenges identified in the preceding regional sections. This section focuses on the special set of challenges facing those living on and associated economically, culturally, and spiritually, with reservations and Native homelands. Although the diversity of land areas and tribal perspectives and situations makes generalizations difficult, a number of key issues illustrating how climate variability and change will affect Native peoples and their communities have been identified.

Observed Climate Trends

Reservations are present in every region of the US, and Native peoples have been experiencing the vagaries of climate on this continent for many thousands of years. Native peoples have developed unique cultures based on the prevailing regional climate, from ice-covered areas of Alaska to the tropical Pacific and Caribbean islands. In each region, however, the climate is starting to change, and Native peoples are aware of these changes. For example, Natives of Alaska are already experiencing significant warming, with the melting of permafrost and sea ice altering subsistence lifestyles (see box on page 74), and changes in the timing of bird and waterfowl migrations as a result of changes in season length are being noticed in many regions.

KEY ISSUES

- Tourism and Community Development

- Human Health and Extreme Events

- Rights to Water and Other Natural Resources

- Subsistence Economies and Cultural Resources

- Cultural Sites, Wildlife, and Natural Resources

The Reality of Living with Ecosystem Shifts

For centuries, the Anishinaabeg (Ojibway or Chippewa) who live around Lake Superior and along the upper Mississippi River have depended upon the natural resources of the forests, lakes, and rivers of the region. Many of the reservation locations were selected to ensure access to culturally sig-

nificant resources, such as maple sugar bushes and wild rice beds, whose locations were thought to be fixed. As drier summer conditions cause the western prairies to shift eastward toward the western Great Lakes, the extents of maple, birch, and wild rice habitats in the US are likely to be significantly reduced. Because Ojibway communities cannot, as a whole, move as ecosystems shift, climate change is likely to

reduce the resources needed to sustain their traditional culture and impact their economic productivity and the value of established treaty rights.

For example, the wild rice that grows abundantly in shallow lake and marshy habitats of northern Wisconsin and Minnesota is likely to be adversely affected. Wild rice plays a critical role in the economic and

Scenarios of Future Climate

Most of the large Indian reservations are located in the central and western US. The Canadian and Hadley model scenarios project warming of as much as 5 to 10°F over the 21st century, with more warming during winters than during summers in many areas. These models also project that, particularly in the Southwest, warmer winters will bring increasing wintertime precipitation, a rising snowline, and earlier springtime runoff, thereby affecting the timing and volume of river flows. Warmer conditions are also projected to lead to increased evaporation, especially in summer, that will dry summer soils and vegetation, more than offsetting the increase in precipitation in some regions. For example, warmer summer conditions are likely to lead to lower river and lake levels in the northern Great Plains and Great Lakes.

Key Issue: Tourism and Community Development

The most urgent priority for tribal governments and communities over the past thirty years has been economic development and job creation. The 1990 census indicated that 31.6% of all Indian people lived below the poverty line, compared to 13.1% of the total population. The sustained growth of the American economy over the past decade has, for the most part, bypassed Indian households and reservations.

Many tribes are basing a significant share of their economic development on recreation and tourism, taking advantage of culturally and historically significant sites and ceremonies and the natural aesthetic beauty of many reservations. These activities provide income while also encouraging the re-establishment of customs and traditions. The economic viability of many of these activities, however, is based on the prevailing climate – water-based recreation on rivers and lakes, forest campsites and trails, and diverse wildlife experiences based on migrating fish and birds and seasonal flowering of plants. As climate changes, these relationships are very likely to change: reduced summer runoff is likely to reduce the flow in many streams, drier summers are likely to increase fire risk and require closure of campgrounds, and the combined effects of climate and ecosystem changes are likely to disrupt wildlife and plant communities.

"We are the ones that live closest to the land, to Mother Earth. We live with it, we experience it, with our hearts and souls, and we depend upon it. When this Earth starts to be destroyed, we feel it."

Caleb Pungowiyi
Yupik Native from Nome, Alaska

For many tribes, particularly those in the Southwest, long-term changes in water resources are likely to have significant consequences for resource-based sectors that depend on stable water supplies.

ceremonial life of many tribes. The hand-harvested and processed seed is highly prized as a gourmet food and adds significant commercial value to the rural reservation economy. Federal treaties guarantee the right of the Anishinaabeg to gather wild rice in their aboriginal territories, which cover much of the states of Wisconsin and Minnesota. As the climate changes, deep or flooding waters in early spring could delay germination of the seed on lake or river bottoms, leading to crop failure. Lower water levels later in the summer could cause the wild rice stalks to break under the weight of the fruithead or make the rice beds inaccessible to harvesters. Extended drought conditions could encourage greater natural competition from more shallow water species. During the dry summer of 1988, conflicts over water pitted federal river management policies against tribal treaty rights and state demands for water.

NATIVE PEOPLES AND HOMELANDS KEY ISSUES

Human Health and Extreme Events

The rural living conditions of many Native Peoples amplify exposure to variations in the weather. Housing on many reservations is old and offers only limited protection from the environment. Although many traditional structures are designed to take advantage of the natural warmth or coolness of the landscape (for example, by being located below ground,having thick walls,or being selectively exposed to or sheltered from the Sun),acclimation,both physiologically and through use of appropriate clothing,is critical, because homes in many areas lack effective heating and cooling systems. While warming in colder regions will relieve some stresses,some acclimation has already occurred. In the presently hot regions,however, there is likely to be a significant increase in stress that will require new responses as new extremes are reached. While an increase in the presence of air-conditioned facilities would help,it would require changes in behavior toward a more indoor lifestyle.

Changes in climate are also likely to create new challenges for community health systems. Drier summer conditions would likely lead to increased lofting of dust and dust-borne organisms and an increase in forest fire incidence. The poorer air quality resulting from increases in smoke and dust could possibly increase respiratory illnesses such as asthma.

Sequences of unusual weather events can also be disruptive. Unusual weather conditions in 1993 led to an outbreak of hantavirus in the Southwest,affecting both human health and perceptions of risk. The infection did not predominantly affect Indian people,but the event caused a significant drop in tourism to southwestern reservations, reducing income for several communities.

Rights to Water and Other Natural Resources

Treaty rights between tribes and the US government provide for allocation of significant amounts of water for use on reservations. As snowmelt and seasonal runoff patterns change,it is possible that water allocations would have to be modified. This would be extremely problematic in the western US if water resources were to become more limited. Because overall precipitation and runoff are projected to rise in some basins,however, it is also possible that water supply problems could be ameliorated if the additional runoff is allocated to users who can accommodate the larger fluctuations in water flows that are projected to occur. For many tribes,particularly those in the Southwest,long-term changes in water resources are likely to have significant consequences for resource-based sectors that depend on stable water supplies.

Subsistence Economies and Cultural Resources

Native lands have provided a wide variety of resources for Native peoples for thousands of years. Forests, grasslands,streams, and coastal zones have provided,and for many groups still provide,substantial amounts of food, fiber,fish,medicines,and culturally important materials. Native traditions are very closely tied to natural events and resources. Although subsistence economies remain a significant basis for family life only in the far north of Canada and Alaska,many tribal communities support themselves by a combination of subsistence, welfare,and market economies.

The subsistence component of Native economies in the Arctic and sub-Arctic is already being threatened by changes in the global climate (see box p.74). Changes in climate,coupled with other human influences, are occurring across the US,and there are projections of more rapid change in the future. In the Plains, warmer winter conditions are already favoring certain types

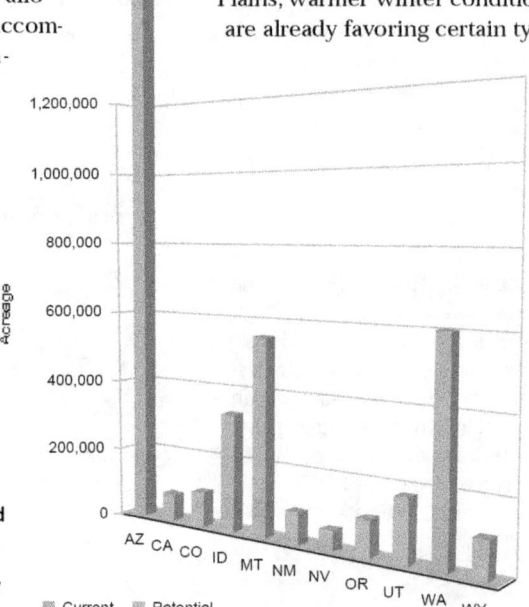

Acreage of Indian lands that are currently being irrigated (brighter blue at bottom of bars) and that could potentially be irrigated (blue at top of bars). Substantially increasing the area of irrigated lands would significantly increase water demand.

of grasses, thereby changing the mix of vegetation types. Shifting away from the subsistence components of their economies is very likely to cause both economic and cultural disruption for many Native peoples.

Cultural Sites, Wildlife, and Natural Resources

The character of local landscapes and weather shapes people's sense of place and how they relate to what surrounds them. While Native peoples have no monopoly on love of land, water, and the sea, their interests started from different premises that have developed over thousands of years of living, moving, and defending their presence on this continent. Although these special connections are frequently explained in spiritual terms, the differences also include intellectual knowledge and historical familiarities extending over thousands of years that continue to be transmitted from generation to generation through oral histories and ceremonies. Many Native peoples perceive humans to be an integral, not dominating, part of the environment.

While there have been significant changes in local environments over past centuries, changes in climate, coupled with other human influences, are likely to bring much larger changes in land cover and wildlife than have occurred in the past. These changes will have practical consequences, but also, at a deeper level, the whole environmental experience that supports religious traditions and the connections to histori-

cally significant sites is likely to start to diverge from what has been sustained through many generations. For Native peoples, externally driven climate change will be disrupting the long history of intimate association with their environments.

Adaptation Strategies

Responding to substantial changes in climate will require technologies and resources, two items desperately scarce in many tribal communities. Most tribal communities are limited in their ways of creating wealth and rely heavily on transfer payments from the federal government. In these communities, adjusting plans for economic and social development to account for climate change may require fresh thinking in federal policies and budgets. In addition, three important steps could be taken.

Enhance Education and Access to Information and Technology

Indian people are significantly underrepresented in scientific and technological professions. They need to develop the understanding and skills to deal with a changing climate. It is especially important to improve the quality of science and technology education in schools and tribal colleges that serve Native youth. It will also be essential to enlist individuals within each Native community to assist in the integration of contemporary information and traditional values.

Promote Local Land-use and Natural Resource Planning

Tribes that have developed strong natural resource management programs for their lands have more substantial bases from which to respond to changes in climate than other tribes. Cost-effective ways, using existing networks and organizations, need to be developed to inform decision-makers in tribal communities, and provide shared access to adequate technical resources.

Participate in Regional and National Discussions and Decision-making

The consequences of changes in climate are rarely contained within reservation boundaries. Serious discussions about climate change must include informed stakeholders from every relevant jurisdiction. A model of interaction and collaboration has been developed between tribes in the northern Great Plains and the University of North Dakota. Their success in broadening participation and making knowledge available in useful ways can provide helpful lessons for other states, tribes, and regions.

SECTOR OVERVIEW

While climate change and variability clearly affect each region quite differently, there are issues of national importance that transcend regional boundaries. Though many such issues were identified, the decision was made to focus on five for this Assessment. These analyses provide a more integrated national picture of the potential consequences of climate variability and change, albeit a picture with regional texture.

These analyses also provide a basis for beginning to think about important interactions between sectors with regard to climate impacts. For example, the projected changes in the timing and amount of precipitation, and hence in water supply, will very likely have significant implications for the other sectors examined here: agriculture, forests, human health, and coastal areas and marine resources. Similarly, the increases in the use of fertilizers and pesticides that are projected for the agricultural sector have obvious implications for all the other sectors as well.

Agriculture

Overall productivity of American agriculture will likely remain high, and is projected to increase throughout the 21st century, with northern regions faring better than southern ones. Though agriculture is highly dependent on climate, it is also highly adaptive. Weather extremes, pests, and weeds will likely present challenges in a changing climate. Falling commodity prices and competitive pressures are likely to stress farmers and rural communities.

Water

Rising temperatures and greater precipitation are likely to lead to more evaporation and greater swings between wet and dry conditions. Changes in the amount and timing of rain, snow, runoff, and soil moisture are very likely. Water management, including pricing and allocation will very likely be important in determining many impacts.

Human Health

Heat-related illnesses and deaths, air pollution, injuries and deaths from extreme weather events, and diseases carried by water, food, insects, ticks, and rodents, have all been raised as concerns for the US in a warmer world. Modern public health efforts will be important in identifying and adapting to these potential impacts.

Coastal Areas and Marine Resources

Coastal wetlands and shorelines are vulnerable to sea-level rise and storm surges, especially when climate impacts are combined with the growing stresses of increasing human population and development. It is likely that coastal communities will be increasingly affected by extreme events. The negative impacts on natural ecosystems are very likely to increase.

Forests

Rising CO_2 concentrations and modest warming are likely to increase forest productivity in many regions. With larger increases in temperature, increased drought is likely to reduce forest productivity in some regions, notably in the Southeast and Northwest. Climate change is likely to cause shifts in species ranges as well as large changes in disturbances such as fire and pests.

AGRICULTURE SECTOR

The US is a major supplier of food and fiber for the world, accounting for more than 25% of the total global trade in wheat, corn, soybeans, and cotton. Cropland currently occupies about 400 million acres, or 17% of the total US land area. In addition, grasslands, and permanent grazing and pasturelands, occupy almost 600 million acres, another 26% of US land area. The value of agricultural commodities (food and fiber) exceeds $165 billion at the farm level and over $500 billion, 10% of GDP, after processing and marketing.

Economic viability and competitiveness are major concerns for producers trying to maintain profitability as real commodity prices have fallen by about two-thirds over the last 50 years. Agricultural productivity has improved at over 1% per year since 1950, resulting in a decline in both production costs and prices. This trend maintains intense pressure on individual producers to continue to increase the productivity of their farms and to reduce costs of production. In this competitive economic environment, producers see anything that might increase costs or limit their markets as a threat to their viability. Issues of concern include regulatory actions that might increase costs, such as efforts to control the off-site consequences of soil erosion, agricultural chemicals, and livestock wastes; growing resistance to and restrictions on the use of genetically modified crops; extreme weather or climate events such as droughts and floods; new pests; and the development of pest resistance to existing pest control strategies. Future changes in climate will interact with all of these factors.

The agriculture sector Assessment considered crop agriculture, grazing, livestock, and environmental effects of agriculture. The focus in this document is primarily on crop agriculture which was studied most intensively in this Assessment. Although extensive, the analysis of crop yields did not fully consider all of the consequences of possible changes in pests, diseases, insects, and extreme events resulting from climate change. This analysis assumes continued technological advances and no changes in federal policies or international trade.

Key Issue: Crop Yield Changes and Associated Economic Consequences

It is likely that climate change, as defined by the scenarios examined in this Assessment, will not imperil the ability of the US to feed its population and to export foodstuffs. Results of this Assessment suggest that, at the national level, productivity of many major crops will likely increase under the climate scenarios used in these crop models. Crops showing generally positive results include cotton, corn for grain and silage, soybeans, sorghum, barley, sugar beets, and citrus fruits. Pastures also show positive results.

KEY ISSUES

- Crop Yield Changes and Associated Economic Consequences

- Changing Water Demands for Irrigation

- Surface Water Quality

- Increasing Pesticide Use

- Climate Variability

Economically, consumers are likely to benefit from lower prices while producers are likely to see their profits decline.

CO_2 Effects on Crops

Greater concentrations of CO_2 generally result in higher photosynthesis rates and may also reduce water losses from plants. Photosynthesis is enhanced when additional carbon is available for assimilation and so crop yields generally rise.

The actual response to increased CO_2 differs among crops. Most commercial crops in the US, including wheat, rice, barley, oats, potatoes, and most vegetable crops, tend to respond favorably to increased CO_2, with a doubling of atmospheric CO_2 concentration leading to yield increases in the range of 15-20%. The crop models used in this Assessment assume a CO_2 fertilization effect in this range, and also assume that sufficient nutrients and water will be available to support these increases. Other crops including corn, sorghum, sugar cane, and many tropical grasses, are less responsive to increases in CO_2, with a doubling of its concentration leading to yield increases of about 5%.

In situations where crop yields are severely limited by factors such as nutrient availability, an enduring CO_2 fertilization effect is very likely to be of only minor importance.

For other crops,including wheat, rice,oats,hay, sugar cane,potatoes,and toma-toes,yields are projected to increase under some conditions and decrease under others. The crop models assume that the CO_2 fertilization effect will be considerable (see box).

In the crop yield models,a limited set of on-farm adaptation options are consid-ered,including changes in planting dates and changes in varieties. These con-tribute small additional gains in yields of dryland crops and greater gains in yields of irrigated crops. The economic models consider a far wider range of adaptations in response to changing productivity, prices,and resource use, including changes in crops and the location of cropping,irrigation,use of fertil-izer and pesticides,and a variety of other farm management options.

All agricultural regions of the US are not affected to the same degree by the cli-mate scenarios studied in this Assessment. In general,this study finds that cli-mate change favors northern areas. The Midwest,West,and Pacific Northwest exhibit large gains in yields with both climate scenarios in the 2030 and 2090 time frames. Crop yield changes in other regions vary more widely depending on the climate scenario and time period. For example,projected wheat yields in western Kansas decline under the Canadian scenario.

...at the national level, pro-ductivity of many major crops will likely increase under the climate-change scenarios used in the crop models. Crops showing generally positive results include cot-ton, corn for grain and silage, soybeans, sorghum, barley, sugar beets, and citrus fruits.

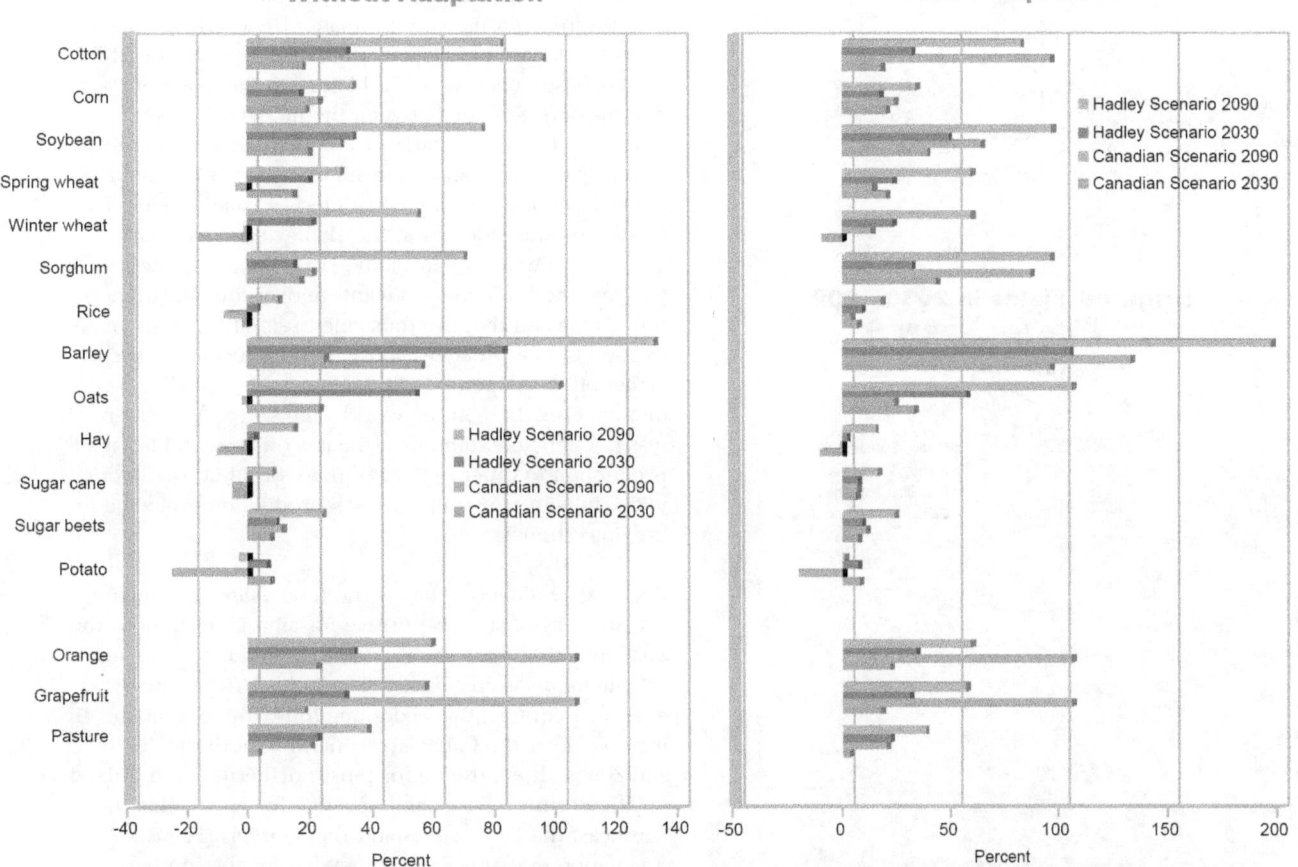

Model simulations of average changes in crop yields for 16 crops. The yield changes are given as percentages and represent the differ-ences between current yields and those projected for two time periods, 2030 and 2090. Two scenarios of future climate, the Canadian and Hadley, were used. The results consider physiological responses of the crops to climate under either dryland or irrigated cultivation. They also consider either "no adaptation" or "adaptation" responses by producers to climate change. Adaptations included changes in planting dates and crop varieties. Only 11 of the 16 crops were actually modeled: cotton, wheat (winter and summer), corn, hay, potato, orange, soybean, sorghum, rice, pasture grass. Results for the other crops are based on extrapolations from the modeled crops.

AGRICULTURE KEY ISSUES

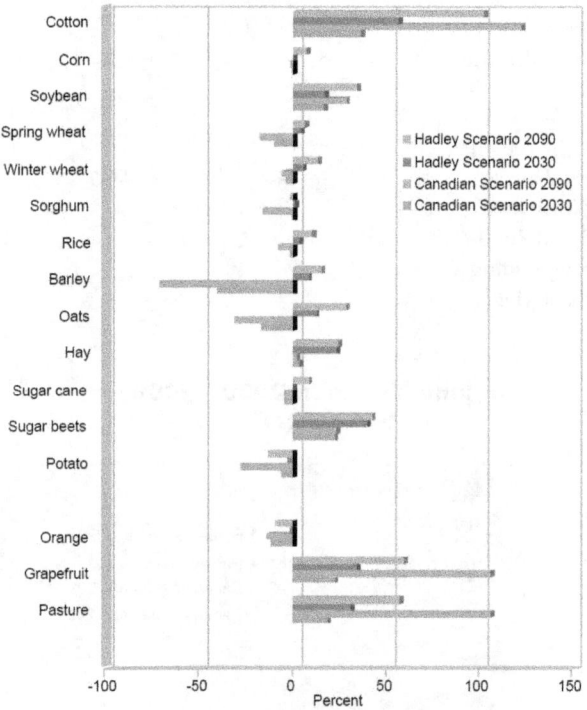

Irrigated Yields in 2030 - 2090
Without Adaptation

Legend:
- Hadley Scenario 2090
- Hadley Scenario 2030
- Canadian Scenario 2090
- Canadian Scenario 2030

Categories (top to bottom): Cotton, Corn, Soybean, Spring wheat, Winter wheat, Sorghum, Rice, Barley, Oats, Hay, Sugar cane, Sugar beets, Potato, Orange, Grapefruit, Pasture

X-axis: Percent (-100, -50, 0, 50, 100, 150)

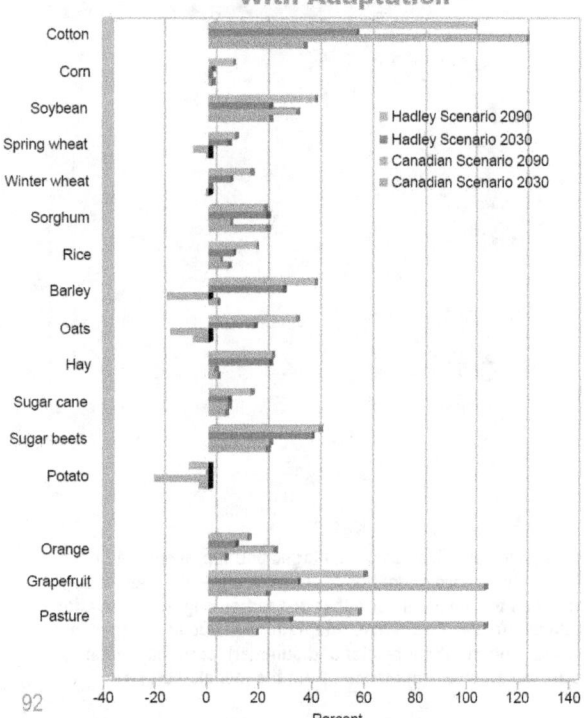

Irrigated Yields in 2030 - 2090
With Adaptation

Legend:
- Hadley Scenario 2090
- Hadley Scenario 2030
- Canadian Scenario 2090
- Canadian Scenario 2030

Categories (top to bottom): Cotton, Corn, Soybean, Spring wheat, Winter wheat, Sorghum, Rice, Barley, Oats, Hay, Sugar cane, Sugar beets, Potato, Orange, Grapefruit, Pasture

X-axis: Percent (-40, -20, 0, 20, 40, 60, 80, 100, 120, 140)

Model simulations suggest that the net effects of the climate scenarios studied on the agricultural segment of the US economy over the 21st century are generally positive. The exceptions are simulations under the Canadian scenario in the 2030 time period, particularly in the absence of adaptation.

Economically, consumers benefit from lower prices while producers' profits decline. Under the Canadian scenario, these opposing economic effects are nearly balanced, resulting in a small net effect on the national economy. The estimated $4-5 billion reduction in producers' profits represents a 13-17% loss of income, while the savings of $3-6 billion to consumers represent less than a 1% reduction in the consumers' food and fiber expenditures. This large difference exists because much of the final cost of agricultural goods to consumers reflects processing, transportation, and retailing costs that the models used here assume are not affected by climate. Under the Hadley scenario, producers' profits decline by up to $3 billion (10%), while consumers save $9-12 billion (in the range of 1%). The major difference between the model outputs is that under the Hadley scenario, productivity increases are substantially greater than under the Canadian, resulting in lower food prices, to the consumers' greater benefit. The smaller producer losses in the Hadley scenario, despite greater productivity gains and price changes, reflect the fact that the US farmers' advantage over foreign competitors grows and they are thus able to significantly increase export volume. Analyses show that producer versus consumer effects depend on how climate change affects production elsewhere in the world. The sector Assessment was not able to extend its estimates on crop and livestock production to other regions of the world but used worldwide shifts in crop and livestock production projected in previous studies.

Regional production change, the total value of crop and livestock production, is positive for all regions in both the 2030 and 2090 time frames under the Hadley scenario. Adaptation measures have a small additional positive effect. In contrast, this economic index differs among regions under the Canadian scenario in both the 2030s and 2090s. It is positive for most northern regions, mixed for the northern Plains, and negative for Appalachia, the Southeast, the Delta states, and the southern Plains. Adaptation measures help somewhat for the southern regions, but the value of production is lower in these regions under both the 2030 and 2090 climates considered.

Changing Water Demands for Irrigation

At the national level, the models used in this Assessment find that irrigated agriculture's need for water declines approximately 5-10% for 2030, and 30-40% for 2090 in the context of the two primary climate scenarios. At least two factors are responsible for this possible reduction. One is increased precipitation in some agricultural areas. The other is that faster development of crops due to higher temperatures results in a reduced growing period and thereby reduced water demand. In the crop modeling analyses done for this Assessment, shortening of the growing period reduces plant water-use enough to more than compensate for the increased water losses from plants and soils due to higher temperatures.

The picture for future agricultural water demands at the regional scale is less clear and it is possible that it will differ substantially from the national picture. At the regional level, there is the possibility that overall water use will increase in response to climate change.

Model simulations suggest that the net effects of climate change on the agricultural segment of the US economy over the 21st century are generally positive.

Regional Production Changes Relative to Current Production
2030 and 2090 periods

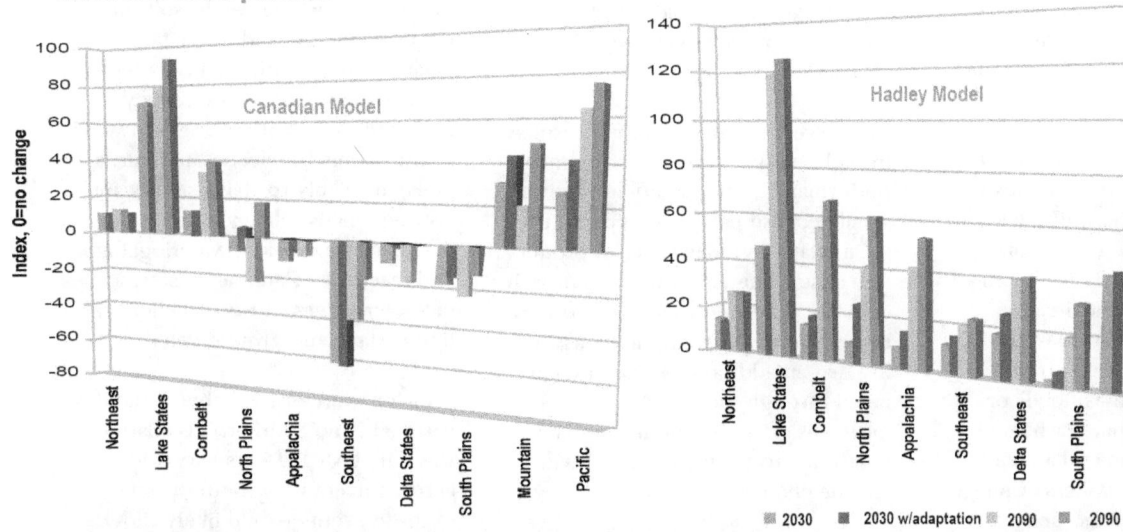

2030   2030 w/adaptation   2090   2090 w/adaptation

Economic Impacts of Climate Change on US Agriculture
2030 and 2090 periods

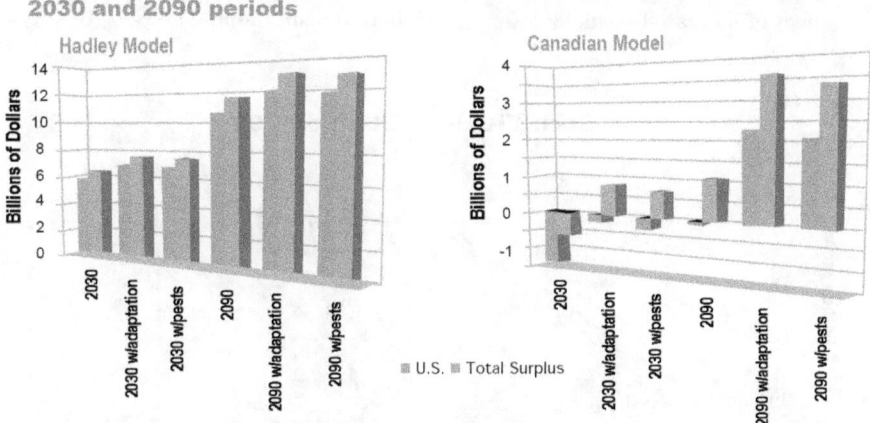

U.S.   Total Surplus

Regional production change (crop and livestock production weighted by prices) from a year 2000 baseline was positive for all regions in both the 2030 and 2090 timeframes under the Hadley scenario. In contrast this index differed among regions under the Canadian scenario in both the 2030s and 2090s. It was positive for most northern regions, mixed for the northern Plains, and negative for Appalachia, the Southeast, the Delta states and the southern Plains.

Economic Impacts of climate change under the Canadian and Hadley climates. The economic index is change in welfare expressed as the sum of producer and consumer surplus in billions of dollars. US (light blue bar above) includes sales and purchases in the US, while Total Surplus (dark blue bar) also includes overseas sales by US producers.

AGRICULTURE KEY ISSUES

Surface Water Quality

A case study of agriculture in the drainage basin of the Chesapeake Bay was undertaken to analyze the effects of climate change on surface-water quality. The Bay is a highly valuable natural resource that has been severely degraded in recent decades. Soil erosion and excess nutrient runoff from crop and livestock production have played a major role in the decline of the Bay's health.

In simulations for this Assessment, under the two climate scenarios for 2030, loading of excess nitrogen into the Chesapeake Bay due to corn production increases by 17-31% compared with the current situation. These projected effects may not fully represent the effects of extreme weather events such as floods or heavy downpours that wash large amounts of fertilizers and animal manure into surface waters. Changes in future farm practices, such as better matching of the timing of plant need for fertilizer with the timing of application, could possibly help to reduce the projected impacts. Because efforts are already underway to protect the Bay, many of these practices may be required and in use before 2030.

Reductions in corn yields often correspond to extreme climate events including droughts and floods. The record Midwest floods of 1993 resulted from this being the wettest year on record, washing out and flooding many corn fields and resulting in late replanting. In 1995, declines in yields resulted from a sequence of unusual climate events; a cool wet spring delayed planting, and a hot, dry summer affected pollination, and ultimately, yield.

Pesticide Use

The Assessment investigates the relationship between pesticide use and climate for crops that require relatively large amounts of pesticide. Pesticide use is projected to increase for most crops studied and in most states, under the climate scenarios considered. Increased need for pesticide application on corn is generally in the range of 10-20%, on potatoes, 5-15%, and on soybeans and cotton, 2-5%. The results for wheat vary widely by state and climate scenario showing changes in pesticide application ranging from approximately –15 to +15%.

The increase in pesticide use results in slightly poorer overall economic performance, but this effect is quite small because pesticide expenditures are a relatively small share of production costs. This Assessment approach does not consider increased crop losses due to pests, implicitly assuming that all additional losses are eliminated through increased pest control measures. This may underestimate losses due to pests associated with climate change.

In addition, this Assessment does not consider the environmental consequences of increased pesticide use

and it is possible that these would be substantial. In a complete economic analysis, the costs of negative impacts of pesticides on the environment would be considered.

Climate Variability

The consequences of climate change for US agriculture are very likely to be affected by changes in climate variability and extreme events. Agricultural systems are vulnerable to climate extremes, with effects varying from place to place because of differences in soils, production systems, and other factors. Changes in precipitation type (rain, snow, or hail), timing, frequency, and intensity, along with changes in wind (windstorms, hurricanes, and tornadoes), are likely to have significant consequences. Heavy precipitation events cause erosion, waterlogging, and leaching of animal wastes, pesticides, fertilizers, and other chemicals into surface and groundwater.

A major source of weather variability is the El Niño Southern Oscillation (ENSO). ENSO effects vary widely across the country. Better prediction of these events would likely allow farmers to plan ahead, altering their choices of which crops to plant and when to plant them. The value of

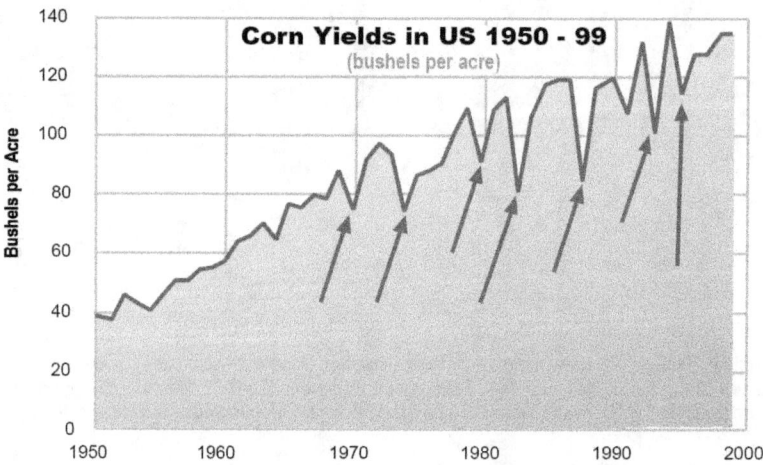

improved forecasts of ENSO events under their current intensity and frequency has been estimated at approximately $500 million per year.

As climate warms, ENSO is likely to be affected. Some models project that more frequent El Niños and stronger La Niñas will have increasing impacts on US weather. The potential impacts of changes in frequency and strength of ENSO conditions on agriculture were modeled in this Assessment. An increase in these conditions is found to cost the US $320 million per year if accurate forecasts of these events are available and farmers use them as they plan for the growing season. The increase in cost is projected to be greater if accurate forecasts are not available or not used.

Adaptation Strategies

Adaptations such as changing planting dates and choosing longer season varieties are likely to offset losses or further increase yields. Adaptive measures are likely to be particularly critical for the Southeast because of the large reductions in yields projected for some crops under the more severe climate scenarios examined. Breeding for response to CO_2 will likely be necessary to achieve the strong fertilization effect assumed in the crop studies. This is an unexploited opportunity and the prospects for selecting for CO_2 response are good. However, attempts to breed for a single characteristic are often not successful, unless other traits and interactions are considered. Breeding for tolerance to climatic stress has already

been heavily exploited and varieties that do best under ideal conditions usually also outperform other varieties under stress conditions. Breeding specific varieties for specific conditions of climate stress is therefore less likely to encounter success.

Some adaptations to climate change and its impacts can have negative secondary effects. For example, an examination of use of water from the Edward's aquifer region around San Antonio, Texas found increased pressure on groundwater resources that would threaten endangered species dependent on spring flows supported by the aquifer. Another example relates to agricultural chemical use. An increase in the use of pesticides and herbicides is one adaptation to increased insects, weeds, and diseases associated with warming. Runoff of these chemicals into prairie wetlands, groundwater, and rivers and lakes could threaten drinking water supplies, coastal waters, recreation areas, and waterfowl habitat.

The wide uncertainties in climate scenarios, regional variation in climate effects, and interactions of environment, economics, and farm policy suggest that there are no simple and widely applicable adaptation prescriptions. Farmers will need to adapt broadly to changing conditions in agriculture, of which changing climate is only one factor. Some of the possible adaptations more directly related to climate include:

- **Sowing dates and other seasonal changes:**
  Plant two crops instead of one or a spring and fall crop with a short fallow period to avoid excessive heat and drought in mid-summer. For already warm growing areas, winter cropping could possibly become more productive than summer cropping.

- **New crop varieties:**
  The genetic base is very broad for many crops, and biotechnology offers new potential for introducing salt tolerance, pest resistance, and general improvements in crop yield and quality.

- **Water supply, irrigation, and drainage systems:**
  Technologies and management methods exist to increase irrigation efficiency and reduce problems of soil degradation, but in many areas, the economic incentives to reduce wasteful practices do not exist. Increased precipitation and more intense precipitation will likely mean that some areas will need to increase their use of drainage systems to avoid flooding and water-logging of soils.

- **Tillage practices:**
  A warmer climate will speed the decay of soil organic matter by bacteria and fungi. Loss of organic matter reduces the capacity of soils to store water and nutrients essential for plant growth. Tillage practices that incorporate crop residues in the soils would likely combat this loss and improve soil quality.

- **Use near-term climate predictions:**
  Accurate six-month to one-year forecasts could possibly reduce losses due to weather variability. For example, predictions of El Niño events have proven useful in regions where El Niño strongly affects weather.

- **Other management adjustments:**
  Virtually all components of the farming system from planting to harvesting to selling might be modified to adjust to climate change.

WATER SECTOR

W ater is a central resource supporting human activities and ecosystems. The hydrologic (water) cycle, a fundamental component of climate, is likely to be altered in important ways by climate change. Precipitation is very likely to continue to increase on average, especially in middle and high latitudes, with much of the increase coming in the form of heavy downpours. Changes in the amount, timing, and distribution of rain, snowfall, and runoff are very probable, leading to changes in water availability as well as in competition for water resources. Changes are also likely in the timing, intensity, and duration of both floods and droughts, with related changes in water quality.

Snowpack serves as natural water storage in mountainous regions and northern portions of the US, gradually releasing its water in spring and summer. Snowpack is very likely to decrease as the climate warms, despite increasing precipitation, for two reasons. It is very likely that more precipitation will fall as rain, and that snowpack will develop later and melt earlier. As a result, peak streamflows will very likely come earlier in the spring, and summer flows will be reduced. Potential impacts of these changes include an increased possibility of flooding in winter and early spring, a reduced possibility of flooding later in the spring, and more shortages in summer.

Managed river systems provide opportunities to store water in reservoirs to dampen the effects of changes in flow regimes, but this does not come without environmental costs. Substantial infrastructure has been developed to store and transport water supplies. There are more than 80,000 dams and reservoirs in the US, and millions of miles of canals, pipes, and tunnels. Even in the absence of climate change, adapting to existing stresses (such as aging infrastructure and inadequate water supplies for growing areas) will be expensive. For a variety of reasons, large dams are no longer viewed as a cost-effective or environmentally acceptable solution to water supply problems, so other strategies must be developed.

KEY ISSUES

- Competition for Water Supplies

- Surface Water Quantity and Quality

- Groundwater Quantity and Quality

- Floods, Droughts, and Extreme Precipitation Events

- Ecosystem Vulnerabilities

Total US Freshwater Consumptive Use

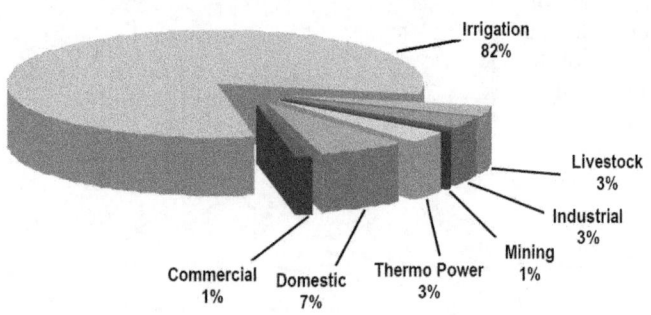

Water Withdrawals and Population Trends

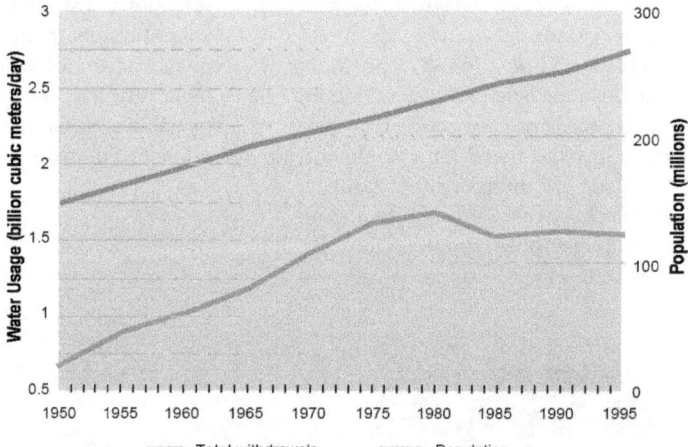

Although US population has continued to increase, total water withdrawals have stablized over the last decade. Stabilization of total withdrawals is due to increased water use efficiency and recycling in some sectors, and a reduction in acreage of irrigated agriculture.

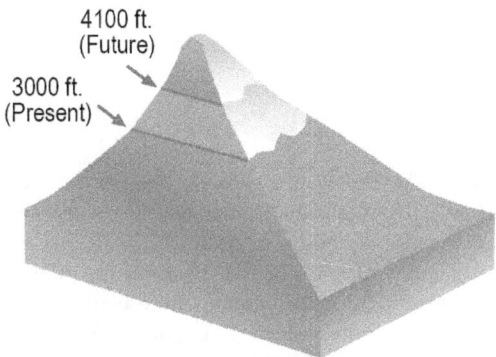

4100 ft.
(Future)

3000 ft.
(Present)

Rough estimate of how much snowlines in the Pacific Northwest
are likely to shift by 2050, assuming about 4°F warming.

Changes in Western Snowpack

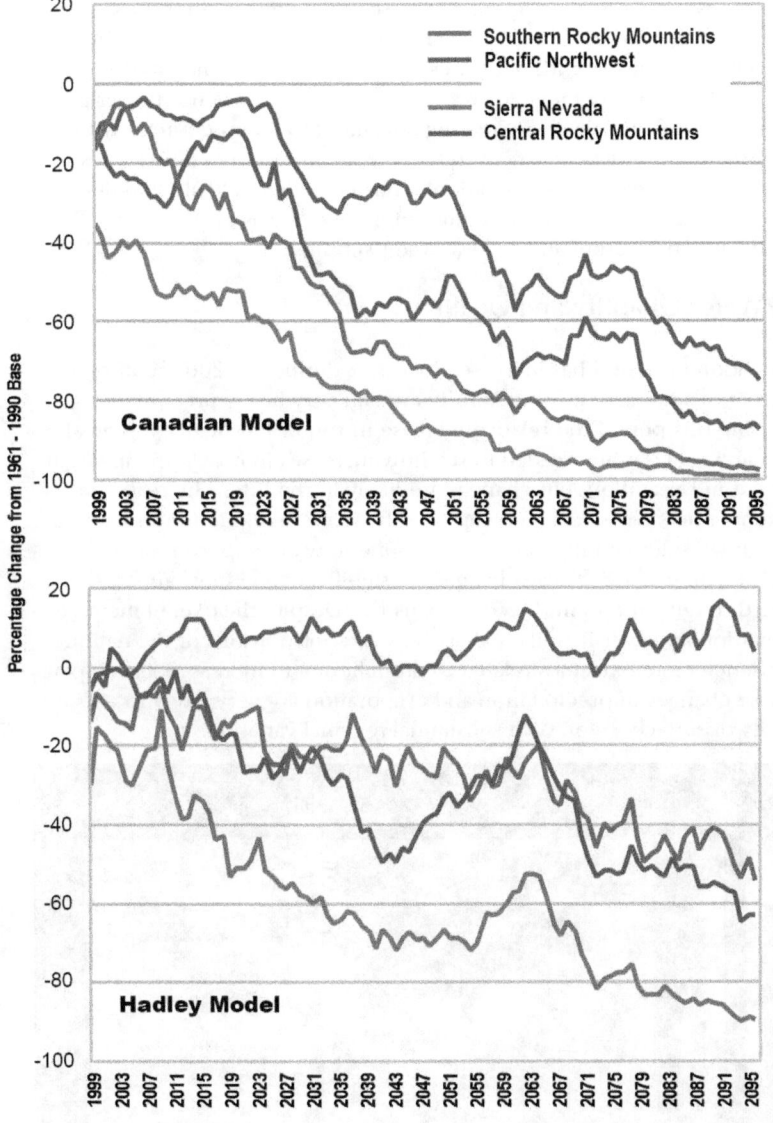

Legend:
— Southern Rocky Mountains
— Pacific Northwest
— Sierra Nevada
— Central Rocky Mountains

Canadian Model

Hadley Model

Percentage Change from 1961 - 1990 Base

Columbia Basin Snow Extent
(Washington & Oregon)

Current

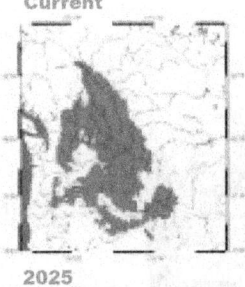

In this model of
Columbia Basin snow
extent, complete loss
of snow cover is pro-
jected at lower eleva-
tions. These maps are
generated by down-
scaling output from
global to regional cli-
mate models for the
Columbia Basin.

2025

2045

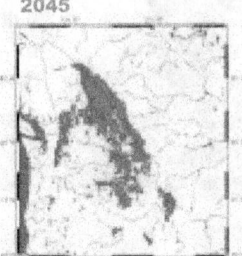

Percentage change from the 1961-90 baseline in the April 1
snowpack in four areas of the western US as simulated for
the 21st century by the Canadian and Hadley models.
April 1 snowpack is important because it stores water that
is released into streams and reservoirs later in the spring
and summer. The sharp reductions are due to rising temperatures and an increasing fraction of winter precipitation
falling as rain rather than snow. The largest changes occur
in the most southern mountain ranges and those closest to
the warming ocean waters.

WATER KEY ISSUES

Competition for Water Supplies

In many rivers and streams in the US, there is not enough water to satisfy existing water rights and claims. Changing public values about preserving in-stream flows, protecting endangered species, and settling Indian water rights claims have made competition for water supplies increasingly intense. Climate change will very likely exacerbate competition in regions where fresh water availability is reduced by increased evaporation due to rising air temperatures and changes in precipitation. In some areas, however, an increase in precipitation could possibly outweigh these factors and increase available supplies.

Significant changes in average temperature, precipitation, and soil moisture caused by climate change are very likely to also affect demand in most sectors, especially in the agriculture, forestry, and municipal sectors. Irrigation water needs are likely to change, with decreases in some places and increases in others. It is very likely that demand for water associated with electric power generation will increase due to the increasing demand for air conditioning with higher temperatures, unless advances in technology make it possible for less water to be used for electrical generation. Climate change is likely to reduce water levels in the Great Lakes and summertime river levels in the central US, thereby affecting navigation and general water supplies.

Surface Water Quantity and Quality

Precipitation in the US has increased by 5-10% during the 20th century with much of this increase attributed to heavy and very heavy precipitation events. During this period, the relative increase in runoff has been even greater. More data and analyses are needed to see how increases in heavy precipitation are reflected in streamflow, but changes are likely in the future. Increases in global temperatures have been accompanied by more precipitation in the middle and high latitudes and increases in atmospheric water vapor in many regions of North America. These changes are significant and most apparent during spring through autumn in the contiguous US. Despite the overall increase in precipitation, however, it is likely that many interior portions of the nation will experience more extremes related to drought due to increased air temperatures. These changes in precipitation and evaporation are very likely to affect the quantity of surface water, with substantial regional variation.

In a warmer climate, hurricanes are likely to produce more rainfall. The frequency and intensity of droughts are also likely to increase in some areas due to higher air temperatures.

The Palmer Drought Severity Index (PDSI) is a commonly used measure of drought severity taking into account differences in temperature, precipitation, and capacity of soils to hold water. These maps show projected changes in the PDSI over the 21st century, based on the Canadian and Hadley climate scenarios. A PDSI of –4 indicates extreme drought conditions. The most intense droughts are in the –6 to –10 range, similar to the major droughts of the 1930s. By the end of the century, the Canadian scenario projects that extreme drought will be a common occurrence over much of the nation, while the Hadley model projects small changes in drought conditions.

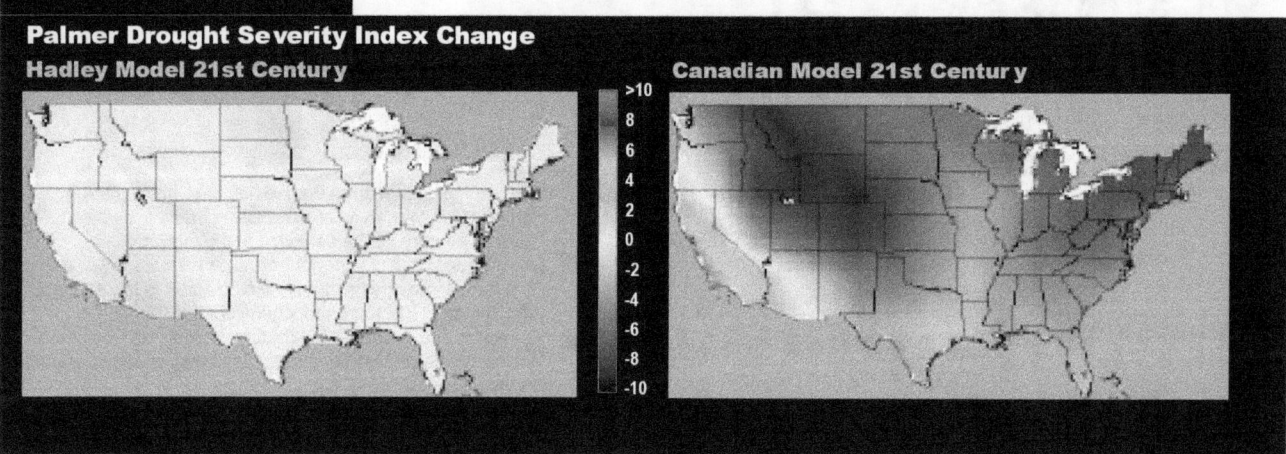

Palmer Drought Severity Index Change

Hadley Model 21st Century

Canadian Model 21st Century

>10
8
6
4
2
0
-2
-4
-6
-8
-10

Rising temperatures are very likely to affect snowfall and increase snowmelt conditions in much of the western and northern portions of the US that depend on winter snowpack for runoff. It is very likely that as the climate warms, less precipitation will fall as snow, the existing snowpack will melt sooner and faster, the runoff will be shifted from late spring and summer to late winter and early spring. This change in the timing of runoff will very likely have implications for water management, flood protection, irrigation, and planning.

Water quality is also likely to be affected by climate change in a variety of ways. For example, more frequent heavy precipitation events will very likely flush more contaminants and sediments into lakes and rivers, degrading water quality. Thus, it is likely that pollution from agricultural chemicals and other non-point sources will be exacerbated. In some regions, however, higher average flows will likely dilute pollutants, improving water quality. Where streamflows are reduced, increased salinity is a potential problem. Water quality issues include potential impacts on human health, such as increased incidence of water-borne diseases after flood events. Flooding can cause overloading of storm and wastewater systems, and damage water and sewage treatment facilities,

mine tailing impoundments, and landfills, thereby increasing the risks of contamination.

Rising water temperatures and changes in ice cover are of particular importance to the ecology of lakes, streams, and their biological communities. Such changes are likely to affect how ecosystems function, especially in combination with chemical pollution. For example, when warmer lake water combines with excess nutrients from agricultural fertilizers (washed into the lake by heavy rains), algae blooms on the lake surface, depleting the ecosystem of oxygen and harming the other organisms in the system.

Groundwater Quantity and Quality

Several regions of the US, including parts of California and the Great Plains region, are dependent on dwindling groundwater supplies. Groundwater supplies are less susceptible than surface water to short-term climate variability; they are more affected by long-term trends. Groundwater serves as the base flow for many streams and rivers. In many areas, groundwater levels are very likely to fall, thus reducing seasonal streamflows. Surface water temperature fluctuates more rapidly with reduced volumes of water, likely

affecting vital habitats. Small streams that are heavily influenced by groundwater are more likely to have reduced streamflows and changes in seasonality of flows, likely damaging existing wetland habitats.

Pumping groundwater at a faster rate than it can be recharged is a major concern, especially in parts of the country that have no other supplies. In the Great Plains, for example, model projections indicate that increased drought conditions are likely, and groundwater levels are already dropping in parts of important aquifers such as the Ogallala.

The quality of groundwater is being diminished by a variety of factors including chemical contamination. Saltwater intrusion is another key groundwater quality concern, particularly in coastal areas where changes in freshwater flows and increases in sea level will both occur. As groundwater pumping increases to serve municipal demand along the coast and less recharge occurs, coastal groundwater aquifers are increasingly affected by seawater. Because the groundwater resource has been compromised by many factors, managers are looking increasingly to surface water supplies which are more sensitive to climate change and variability.

Observed Changes In Streamflow and Precipitation (1939-99)

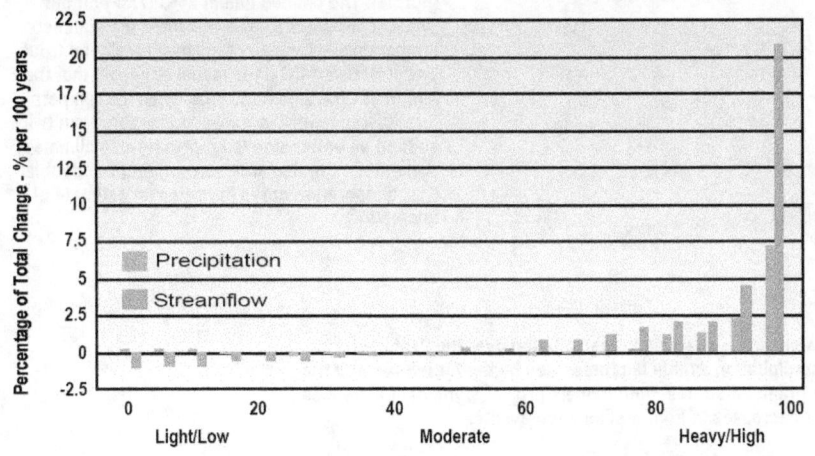

The graph shows changes in the intensity of precipitation and streamflow, displayed in 5% increments, during the period 1939-99 based on over 150 unregulated streams across the US with nearby precipitation measurements (with most of the co-located gages from the eastern half of the US). As the graph demonstrates, the largest changes have been the significant increases in the heaviest precipitation events and the highest streamflows. Note that changes in streamflow follow changes in precipitation, but are amplified by about a factor of 3.

WATER KEY ISSUES

Floods, Droughts, and Extreme Precipitation Events

Changes in climate extremes are more likely to cause stress at the regional level than changes in the averages. Thus, changes in the timing of precipitation events, as well as increases in extreme events, are key concerns. Climate change is likely to increase flood frequency and amplitude in some regions, with major impacts on infrastructure and emergency management. The 1999 North Carolina flood, resulting from Hurricane Floyd, offers a recent example of the massive dislocations and multi-billion dollar costs that often accompany such events. In a warmer climate, hurricanes are likely to produce more rainfall. The frequency and intensity of droughts are also likely to increase in some areas due to higher air temperatures. This is likely to have wide-ranging impacts on agriculture, water-based transportation, and ecosystems. A recent example of such impacts is the 1995-96 drought in the agricultural regions of the southern Great Plains that resulted in about $5 billion in damages. The costs associated with floods and droughts include those incurred for building and managing infrastructure to avoid damages as well as costs associated with damages that are not avoided. These costs are in the billions of dollars and rising.

Ecosystem Vulnerabilities

Species live in the larger context of ecosystems and have differing environmental needs. A change that is devastating to one species is likely to encourage the expansion of another to fill that niche in the system. Extreme conditions such as floods, droughts, and fire are critical to sustaining certain ecosystems, and changes in the frequency of these events are likely.

Rising temperatures in surface waters are likely to force out some cold water fish species such as salmon and trout that are already near the threshold of their viable habitat. Increasing temperatures also result in reduced dissolved oxygen in water, reducing ecosystem health. Temperature increases will very likely reduce ice cover and alter mixing and stratifica-

Summer Stream Temperatures
Steamboat Creek, Oregon

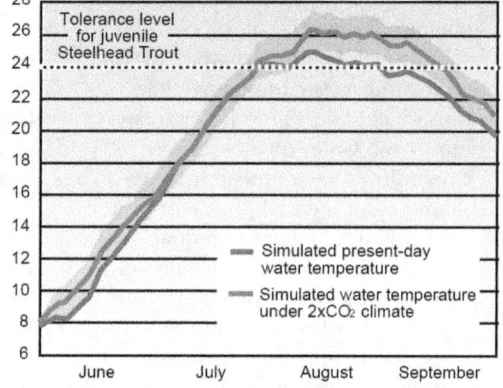

Simulated summer stream temperatures under present day climate (blue) and simulated temperatures under about a twice current CO_2 climate (purple). The dashed line at 24°C (75°F) on the "water temperature" axis indicates the summer temperature tolerance of juvenile steelhead trout. Under doubled CO_2, the model suggests that the length of time within the year when the temperature tolerance limit is exceeded is more than twice as long as under simulated present-day climate conditions. Shaded area surrounding the doubled CO_2 temperature curve indicates an estimate of uncertainty.

Projected 21st Century Change in US Daily Precipitation

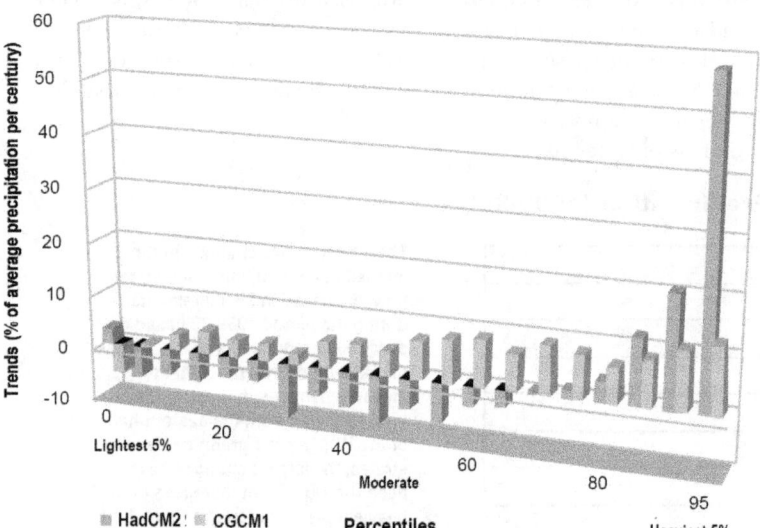

These projections by the Hadley and Canadian models show the changes in precipitation over the 21st century. Each models' projected change in the lightest 5% of precipitation events is represented by the far left bar and the change in the heaviest 5% by the far right bar. As the graph illustrates, both models project significant increases in heavy precipitation events with smaller increases or decreases in light precipitation events.

tion of water in lakes, all of which are key to the nutrient balance and habitat value. The natural ecosystems of the Arctic, Great Lakes, Great Basin, Southeast, and the prairie potholes of the Great Plains appear highly vulnerable to the projected changes in climate. In regions where runoff increases, existing stresses and threats to biodiversity could possibly be reduced.

Adaptation Strategies

Strategies for adapting to climate change range from changes in the operation of dams and reservoirs, to re-evaluating basic engineering assumptions used in facility construction, to building new infrastructure. Options also include water conservation, use of reclaimed wastewater, water transfers, and increasing prices (which encourages increases in efficiency of use). Because many of the impacts of climate change are not predictable, more flexible institutional arrangements are needed in order to adapt to changing conditions including not only climate change, but other existing stresses as well. Water rights systems vary from state to state, with even more differences at the local level. Most institutions related to water have not responded well to changing socioeconomic and environmental conditions.

Some have argued that an open market in water rights would help resolve conflict and increase efficiency because water would flow to the highest and best use based on willingness to pay. Although major social, equity, and environmental considerations must be addressed, market solutions appear to have great potential to help resolve supply problems in some parts of the US.

In considering adaptation mechanisms, it is important to point out that humans have a great ability to adapt to change, while natural ecosystems are likely to be more vulnerable. Some potential adaptation options for human water management in response to climate change and other stresses follow.

- Increase ability to shift water within and between sectors (including agriculture to urban).

- Use pricing and market mechanisms proactively to increase efficiency of water use.

- Incorporate potential changes in demand and supply in long-term planning and infrastructure design.

- Create incentives or requirements to move people and structures out of floodplains.

- Identify ways to manage all available supplies, including groundwater, surface water, and effluent, in a sustainable manner.

- Restore and maintain watersheds (for example, by restoring appropriate vegetation) as an integrated strategy for managing water quality and quantity. Restoring watersheds that have been damaged by urbanization, forestry, or grazing can reduce sediment loads and nutrients in runoff, limit flooding, and reduce water temperature.

- Reuse municipal wastewater, improve management of urban storm water runoff, and promote collection of rainwater for local use to enhance urban water supplies.

- Increase the use of forecasting tools for water management. Some weather patterns, such as those resulting from El Niños, can now be predicted with some accuracy, and this can help reduce damages associated with extreme events.

- Enhance monitoring efforts to improve data for weather, climate, and hydrologic modeling to aid understanding of water-related impacts and management strategies.

Prairie pothole.

The natural ecosystems of the Arctic, Great Lakes, Great Basin, Southeast, and the prairie potholes of the Great Plains appear highly vulnerable to the projected changes in climate.

Land subsidence, caused by over-pumping of groundwater, can result in earth fissures such as this near Eloy, Arizona.

Artificial groundwater recharge in the Santa Ana Riverbed, Orange County Water District, California.

The Central Arizona Project brings Colorado River water 330 miles uphill to Tucson and Phoenix, Arizona.

HEALTH SECTOR

C ertain health outcomes are known to be associated with weather and/or climate, including:illnesses and deaths associated with temperature;extreme precipitation events;air pollution;water contamination;and diseases carried by mosquitoes,ticks,and rodents. Because human health is intricately bound to weather and the many complex natural systems it affects,it is possible that projected climate change will have measurable impacts,both beneficial and adverse,on health. Projections of the extent and direction of potential impacts of climate variability and change on health are extremely difficult to make because of many confounding and poorly understood factors associated with potential health outcomes,population vulnerability, and adaptation. For example,not enough is yet known about particulate matter to project how levels of this air pollutant might change in projected future climate scenarios. Basic information on the sensitivity of human health to aspects of weather and climate is limited,and it is difficult to anticipate what adaptive measures might be taken in the future to mitigate risks of adverse health outcomes,such as vaccines or improved use of weather forecasting.

Health outcomes in response to climate change are highly uncertain. Currently available information suggests that a range of negative health impacts is possible. These have been the focus of much of the public health research on climate change to date. Some positive health outcomes,notably reduced cold-weather mortality, are possible,although the balance between increased risk of heat-related illnesses and death and changes in winter illnesses and death cannot yet be confidently assessed. At present, much of the US population is protected against adverse health outcomes associated with weather and/or climate, although certain demographic and geographic populations are at greater risk. Adaptation, primarily through the maintenance and improvement of public health systems and their responsiveness to changing climate conditions and to identified vulnerable subpopulations should help to protect the US population from adverse health outcomes of projected climate change. The costs,benefits,and availability of resources for such adaptation must be considered,and further research into key knowledge gaps on the relationships between climate/weather and health is needed.

Temperature-related Illnesses and Deaths

E pisodes of extreme heat already pose a health threat in parts of the US. For example, following a five-day heat wave in 1995 in which maximum temperatures in Chicago,

KEY ISSUES

- Temperature-related Illnesses and Deaths

- Health Effects Related to Extreme Weather Events

- Air Pollution-related Health Effects

- Water- and Food-borne Diseases

- Insect-, Tick- and Rodent-borne Diseases

July Heat Index Change - 21st century

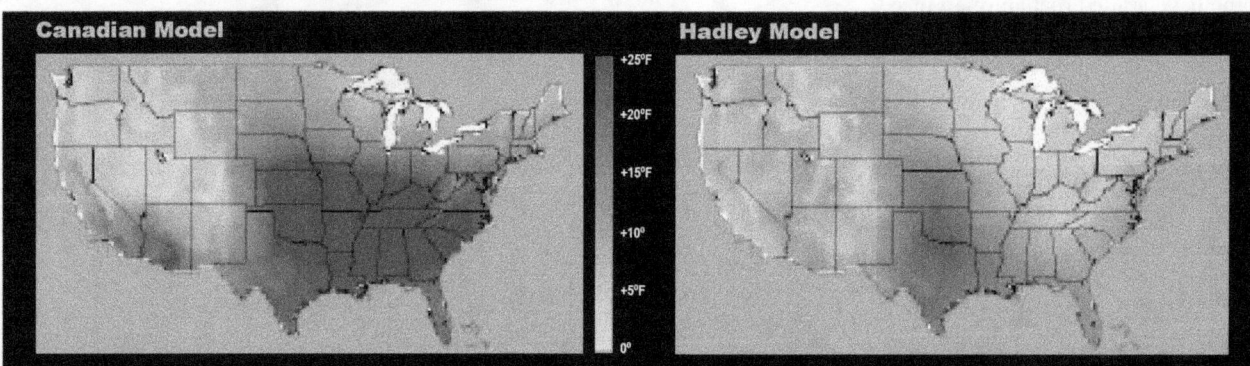

Both models project substantial increases in the July heat index (which combines heat and humidity) over the 21st Century. These maps show the projected increase in average daily July heat index relative to the present. The largest increases are in the southeastern states, where the Canadian model projects increases of more than 25˚F. For example, a July day in Atlanta that now reaches a heat index of 105˚F would reach a heat index of 115˚F in the Hadley model, and 130˚F in the Canadian model.

Illinois ranged from 93 to 104°F, the number of deaths increased 85% over the number recorded during the same period of the preceding year. At least 700 excess deaths (deaths in that population beyond those expected for that period of time) were recorded,most of which were directly attributable to heat. Studies in certain urban areas show a strong association between increases in mortality and increases in heat,measured by maximum or minimum daily temperature and heat index (a measure of temperature and humidity).Some of these studies adjust for other weather conditions.

Heat stroke and other health effects associated with exposure to extreme and prolonged heat appear to be related to environmental temperatures above those to which the population is accustomed. Thus,the regions most sensitive to projected increases in severity and frequency of heatwaves are likely to be those in which extremely high temperatures occur only irregularly.

Within heat-sensitive regions,populations in urban areas are most vulnerable to adverse heat-related health outcomes.Heat indices and heat-related mortality rates are higher in the urban core than in surrounding areas. Urban areas remain warmer throughout the night compared to outlying suburban and rural areas. The absence of nighttime relief from heat for urban residents is a factor in excessive heat-related deaths. The elderly, young children,the poor, and people who are bedridden,on certain medications,or who have certain underlying medical conditions are at particular risk.

Overall death rates are higher in winter than in summer, and it is possible that milder winters could reduce deaths in winter months. However, the relationship between winter weather and mortality is difficult to interpret. For example,many winter deaths are due to respiratory infections such as influenza,and it is unclear how influenza transmission would be affected by higher winter temperatures. The net effect on winter mortality from climate change is therefore extremely uncertain.

...following a five-day heat wave in 1995 in which maximum temperatures in Chicago, Illinois ranged from 93° to 104°F, the number of deaths increased 85% over the number recorded during the same period of the preceding year.

Deaths due to summer heat are projected to increase in US cities, in a study using several climate models. Mortality rates (number of deaths per 100,000 population) are shown for the Max-Planck Institute model, the results from which lie roughly in the middle of the models examined. Because heat-related illness and death appear to be related to temperatures much hotter than those to which the population is accustomed, cities that experience extreme heat only infrequently appear to be at greatest risk. For example, Philadelphia, New York, Chicago, and St. Louis have experienced heat waves that resulted in a large number of heat-related deaths, while heat-related deaths in Atlanta and Los Angeles are much lower. In this study, statistical relationships between heat waves and increased death rates are constructed for each city based on historical experience. Deaths under a city's future climate are then projected by applying that city's projected incidence of extreme heat waves to the statistical relationship that was estimated for the city whose present climate is most similar to the projected future climate for the city in question. This approach attempts to represent how people will acclimate to the new average climate they experience.

Heat Related Deaths - Chicago, July 1995
Maximum Temperature and Heat Index

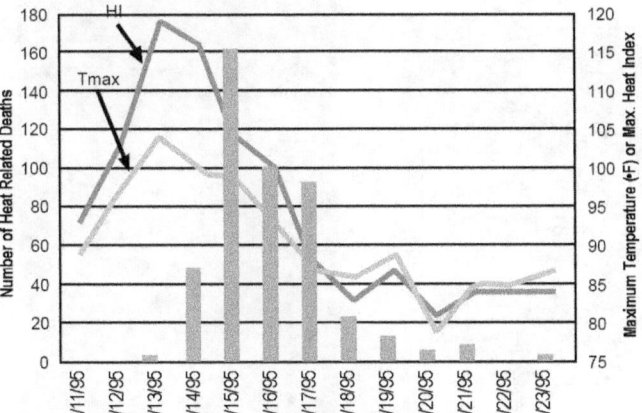

This graph tracks maximum temperature (Tmax), heat index (HI), and heat-related deaths in Chicago each day from July 11 to 23, 1995. The gray line shows maximum daily temperature, the blue line shows the heat index, and the bars indicate number of deaths for the day.

Average Summer Mortality Rates
Attributed to hot weather episodes

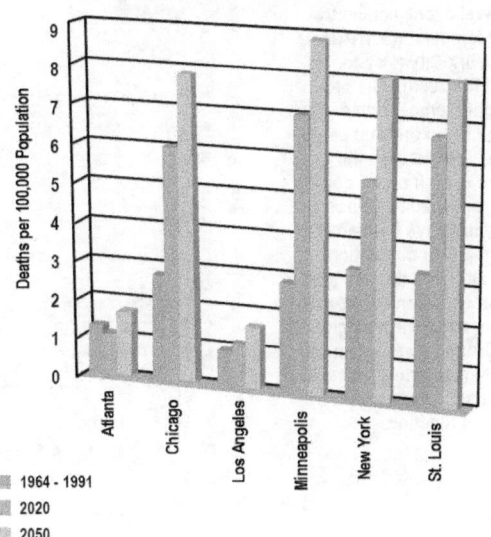

■ 1964 - 1991
■ 2020
■ 2050

HEALTH KEY ISSUES

Changes in precipitation, temperature, humidity, salinity, and wind have a measurable effect on water quality. In 1993, the Milwaukee, Wisconsin drinking water supply became contaminated by Cryptosporidium, and as a result 400,000 people became ill and 54 died.

Heat and heat waves are very likely to increase in severity and frequency with increasing global average temperatures. The climate scenarios used in this Assessment show increases in average summer temperatures and relatively larger increases in average winter temperatures,leading to new record high temperatures, both in summer and winter. The size of US cities and the proportion of US residents living in them are also projected to increase over the next century, so it is possible that the population at risk from heat events will increase.

Heat-related illnesses and deaths are largely preventable through behavioral adaptations,including use of air conditioning and increased fluid intake. However, the degree to which these adaptations might be broadly adopted or economically available to sensitive populations has not been assessed.

Health Effects Related to Extreme Weather Events

Injury and death are the direct health impacts most often associated with natural disasters such as floods and hurricanes.Secondary health effects have also been observed. These effects are mediated by changes in ecological systems (such as bacterial and fungal proliferation) and in public health infrastructures (such as the availability of safe drinking water). The health impacts of extreme weather events such as floods and storms therefore hinge on the vulnerabilities and recovery capacities of the natural environment and the local population. There is controversy about the incidence and continuation of significant mental problems,such as post traumatic stress disorder, following disasters. However, a rise in mental disorders has been observed following several natural disasters in the US.

These graphs illustrate the observed association between ground-level ozone concentrations and temperature in Atlanta and New York City (May to October 1988-1990). The projected higher temperatures across the US in the 21st century are likely to increase the occurence of high ozone concentrations, especially since extremely hot days frequently have stagnant air circulation patterns, although this will also depend on emissions of ozone precursors and meteorological factors. Ground-level ozone can exacerbate respiratory diseases and cause short-term reductions in lung function.

Maximum Daily Ozone Concentrations
and Maximum Daily Temperature - Atlanta & New York

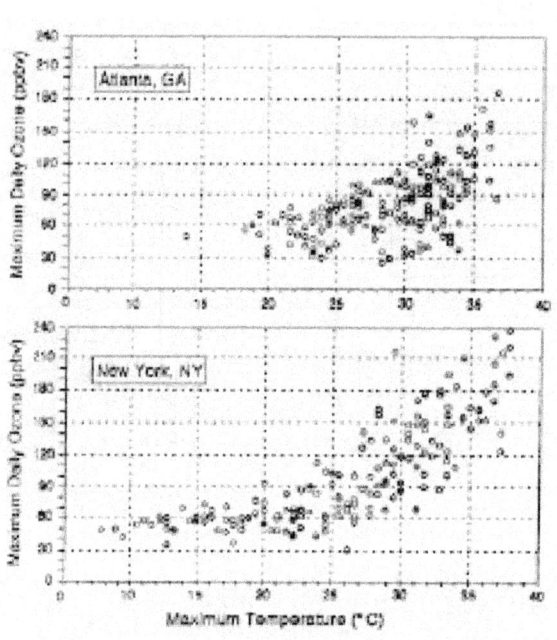

Atlanta

104

Increases in heavy precipitation have occurred in the US over the past century. Future climate scenarios show likely increases in the frequency of extreme precipitation events, including precipitation during hurricanes. This poses an increased risk of floods and associated health impacts.

Air Pollution-related Health Effects

Current exposures to air pollution have serious public health consequences. Ground-level ozone can exacerbate respiratory diseases and cause short-term reductions in lung function. Exposure to particulate matter can aggravate existing respiratory and cardiovascular diseases, alter the body's defense systems against foreign materials, damage lung tissue, lead to premature death, and possibly contribute to cancer. Health effects of exposure to carbon monoxide, sulfur dioxide, and nitrogen dioxide can include reduced work capacity, aggravation of existing cardiovascular diseases, effects on breathing, respiratory illnesses, lung irritation, and alterations in the lung's defense systems.

The mechanisms by which climate change affects exposures to air pollutants include 1) affecting weather and thereby local and regional pollution concentrations; 2) affecting human-caused emissions, including adaptive responses involving increased fuel combustion for power generation; 3) affecting natural sources of air pollutant emissions; and 4) changing the distribution and types of airborne allergens. Analyses show that higher surface air temperatures are conducive to increased concentrations of ground-level ozone. Since it is very likely that temperatures will increase significantly across the US by the end of the 21st century, this creates a risk of higher concentrations of ground-level ozone, especially because higher temperatures are frequently accompanied by stagnating circulation patterns. However, without knowledge of future emissions in specific places, the success of air pollution policies, and local and regional meteorological scenarios, more specific predictions of exposure to air pollutants and health effects cannot be made with confidence.

In addition to affecting exposure to air pollutants, there is some chance that climate change will play a role in exposure to airborne allergens. Climate change will possibly alter pollen production in some plants and the geographic distribution of plant species. Consequently, there is some chance that climate change will affect the timing or duration of seasonal allergies. The impact of pollen and of pollen changes on the occurrence and severity of asthma, the most common chronic disease of childhood, is currently very uncertain.

Water- and Food-borne Diseases

Exposure to water-borne disease can result from drinking contaminated water, eating seafood from contaminated water, eating fresh produce irrigated or processed with contaminated water, or from activities such as fishing or swimming in contaminated water. Water-borne pathogens of current concern include viruses, bacteria (such as *Vibrio vulnificus*, a naturally-occurring estuarine bacterium responsible for a high percentage of the deaths associated with shellfish consumption), and protozoa (such as *Cryptosporidium*, associated with gastrointestinal illnesses). Changes in precipitation, temperature, humidity, salinity, and wind have a measurable effect on water quality. In 1993, the Milwaukee, Wisconsin drinking water supply became contaminated by *Cryptosporidium*, and as a result 400,000 people became ill. Of the 54 individuals who died, most had compromised immune systems because of HIV infection or other illness. A contributing factor in the contamination, in addition to treat-

Injury and death are the direct health impacts most often associated with natural disasters such as floods and hurricanes. Future climate scenarios show likely increases in the frequency of extreme precipitation events, including precipitation during hurricanes. This poses an increased risk of floods and associated health impacts.

Since it is very likely that temperatures will increase significantly across the US by the end of the 21st century, this creates a risk of higher concentrations of ground-level ozone, especially since higher temperatures are frequently accompanied by stagnating circulation patterns.

HEALTH KEY ISSUES

ment system malfunctions, was heavy rainfall and runoff that resulted in a decline in the quality of raw surface water arriving at the Milwaukee drinking water plants. In Florida during the strong El Niño winter of 1997-1998, heavy precipitation and runoff greatly elevated the counts of fecal bacteria and infectious viruses in local coastal waters. In Gulf Coast waters, *Vibrio vulnificus* bacteria are especially sensitive to water temperature, which dictates their seasonality and geographic distribution. In addition, toxic red tides proliferate as seawater temperatures increase. Reports

of marine-related illnesses have risen over the past two and a half decades along the East Coast, in correlation with El Niño events.

Climate changes projected to occur in the next several decades, in particular the likely increase in extreme precipitation events, will probably raise the risk of contamination events.

Insect-, Tick-, and Rodent-borne Diseases

Malaria, yellow fever, dengue fever, and other diseases transmitted between humans by blood-feeding insects, ticks, and mites were once common in the US. Many of these diseases are no longer present, mainly because of changes in land use, agricultural methods, residential patterns, human behavior, and vector

control. However, diseases that may be transmitted to humans from wild animals continue to circulate in nature in many parts of the country. Humans may become infected with the pathogens that cause these diseases through transmission by insects or ticks (such as Lyme disease, which is tick-borne) or by direct contact with the host animals or their body fluids (such as hantaviruses, which are carried by numerous rodent species and transmitted to humans through contact with rodent urine, droppings, and saliva). The organisms that directly transmit these diseases are known as vectors.

The ecology and transmission dynamics of these vector-borne infections are complex, and the factors that influence transmission are unique for each pathogen. Most vector-borne diseases exhibit a distinct seasonal

Combined Wastewater Systems

Wastewater systems that combine storm drains, sewage and industrial waste are still used in about 950 communities, mostly in the Northeast and Great Lakes regions. During rainstorms or spring snowmelt, when the volume of water being discharged can exceed the capacity of the sewage treatment system, these systems are designed to overflow and discharge untreated sewage into surface waters. In 1994, EPA developed a framework to control such combined-sewer overflows under the federal Clean Water Act's water discharge permit program. If combined sewer systems remain in place and continue to discharge untreated wastewater during storms, they will very likely pose an increased health risk under projected increases in intense precipitation events.

Reported Cases of Dengue 1980-1996

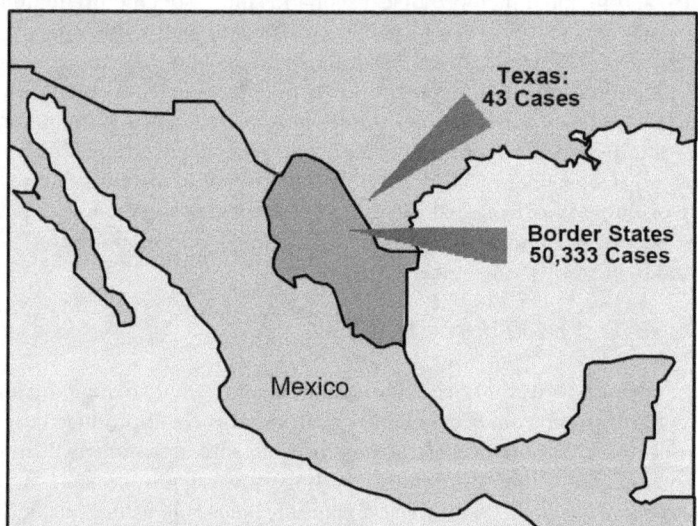

Dengue along the US-Mexico border. Dengue, a mosquito-borne viral disease, was once common in Texas (where there were an estimated 500,000 cases in 1922), and the mosquito that transmits it remains abundant. The striking contrast in the incidence of dengue in Texas versus three Mexican states that border Texas (43 cases vs. 50,333) in the period from 1980-1996 provides a graphic illustration of the importance of factors other than temperature, such as public health infrastructure, use of air conditioning and window screens, in the transmission of vector-borne diseases.

pattern, which clearly suggests that they are weather sensitive. Rainfall, temperature, and other weather variables affect both vectors and the pathogens they transmit in many ways. For example, epidemics of malaria are associated with rainy periods in some parts of the world, but with drought in others. Higher temperatures may increase or reduce vector survival rate, depending on each specific vector, its behavior, ecology, and many other factors.

In some cases, specific weather patterns over several seasons appear to be associated with increased transmission rates. For example, in the midwestern US, outbreaks of St. Louis encephalitis (a viral infection of birds that can also infect and cause disease in humans) appear to be associated with the sequence of warm, wet winters, cold springs, and hot dry summers. The factors underlying this association are complex and require more investigation.

Adaptation Strategies

The future vulnerability of the US population to the health impacts of climate change largely depends on the magnitude of the increase in potential health impacts and on our capacity to adapt to potential adverse changes through legislative, administrative, institutional, technological, educational, and research-related measures. Examples include building codes and zoning to prevent storm or flood damage, severe weather warning systems, improved disease surveillance and prevention programs, improved sanitation systems, education of health professionals and the public, and research addressing key knowledge gaps in climate/health relationships.

Many of these adaptive responses are desirable from a public health perspective irrespective of climate change. For example, reducing air pollution obviously has both short- and long-term health benefits.

Improving warning systems for extreme weather events and eliminating existing combined sewer and storm water drainage systems are other measures that can ameliorate some of the potential adverse impacts of current climate extremes and of the possible impacts of climate change. Improved disease surveillance, prevention systems, and other public health infrastructure at the state and local levels are already needed. Adaptation is a complex undertaking, as demonstrated by the varying degrees of effectiveness of current efforts to cope with climate variability. Considerable work still needs to be done to assess the feasibility (for example, the ability of a community to incur the costs) and the effectiveness of alternative adaptive responses, and to develop improved mechanisms for coping with climate variability and change.

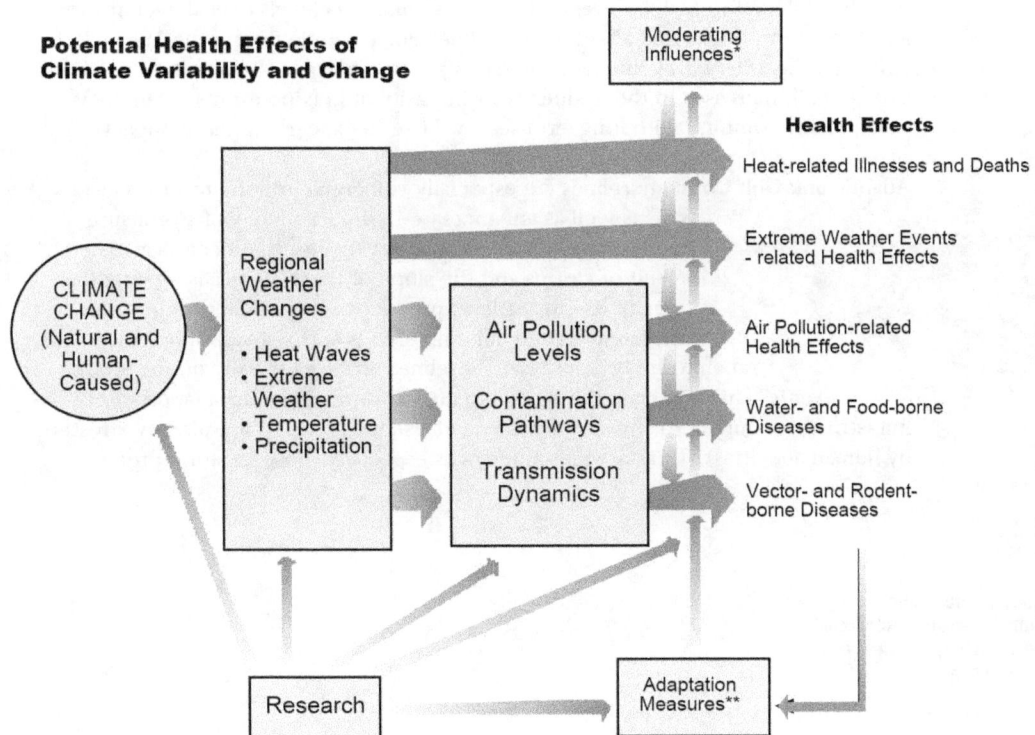

Potential Health Effects of Climate Variability and Change

Health Effects
- Heat-related Illnesses and Deaths
- Extreme Weather Events - related Health Effects
- Air Pollution-related Health Effects
- Water- and Food-borne Diseases
- Vector- and Rodent-borne Diseases

CLIMATE CHANGE (Natural and Human-Caused)

Regional Weather Changes
- Heat Waves
- Extreme Weather
- Temperature
- Precipitation

Air Pollution Levels

Contamination Pathways

Transmission Dynamics

Moderating Influences*

Research

Adaptation Measures**

*Moderating influences include non-climate factors that affect climate-related health outcomes, such as population growth and demographic change, standards of living, access to health care, improvements in health care, and public health infrastructure.

**Adaptation measures include actions to reduce risks of adverse health outcomes, such as vaccination programs, disease surveillance, monitoring, use of protective technologies, such as air conditioning, pesticides, water filtration/treatment, use of climate forecasts and development of weather warning systems, emergency management and disaster preparedness programs, and public education.

COASTAL AREAS AND MARINE RESOURCES

The US has over 95,000 miles of coastline and over 3.4 million square miles of ocean within its territorial waters. These areas provide a wide range of essential goods and services to society. Some 53% of the total US population live on the 17% of land in the coastal zone, and these areas become more crowded every year. Because of this growth, as well as increased wealth and affluence, demands on coastal and marine resources for both aesthetic enjoyment and economic benefits are rapidly increasing.

Coastal and marine environments are intrinsically linked to climate in many ways. The ocean is an important distributor of the planet's heat, with major ocean currents moving heat toward the poles from the equator. There is some chance that this distribution of heat through the ocean's "conveyor belt" circulation would be strongly influenced by the changes projected in many global climate models. Sea-level rise is another climate-related phenomenon with a major influence on coastlines. Global sea level has already risen by 4 to 8 inches (10-20 cm) in the past century and models suggest this rise is very likely to accelerate. The best estimate is that sea level will rise by an additional 19 inches (48 cm) by 2100, with an uncertainty range of 5 to 37 inches (13-95 cm). Geological forces (such as subsidence, in which the land falls relative to sea level) play a prominent role in regional sea-level change. Accelerated global sea-level rise is expected to have dramatic impacts in those regions where subsidence and erosion problems already exist.

KEY ISSUES

- Shoreline Erosion and Human Communities

- Threats to Estuarine Health

- Coastal Wetland Survival

- Coral Reef Die-offs

- Stresses on Marine Fisheries

Key Issue: Shoreline Erosion and Human Communities

Coastal erosion is already a widespread problem in much of the country and has significant impacts on undeveloped shorelines as well as on coastal development and infrastructure. Along the Pacific Coast, cycles of beach and cliff erosion have been linked to El Niño events that raise average sea levels over the short term and alter storm tracks that affect the coastline. For example, during the 1982-83 El Niño and the 1997-98 El Niño, erosion damage was widespread along the Pacific Coastline. If increases in the frequency or intensity of El Niño events occur, they would likely combine with long-term sea-level rise to exacerbate these impacts.

Atlantic and Gulf Coast shorelines are especially vulnerable to long term sea-level rise as well as any increase in the frequency of storm surges or hurricanes. Most erosion events on these coasts are the result of storms, and the slope of these areas is so gentle that a small rise in sea level produces a large inland shift of the shoreline. When buildings, roads, and seawalls block this natural shift, the beaches and shorelines erode, especially during storm events. This increases the threats to coastal development, transportation infrastructure, tourism, freshwater aquifers, and fisheries (which are already stressed by human activities). Coastal cities and towns, especially those in storm-prone

Coastal Vulnerabilty

- Low
- Moderate
- High
- Very High

The preliminary results shown on this map illustrate the relative vulnerability to sea-level rise along the New York and New Jersey coastline.

regions such as the Southeast, are particularly vulnerable to extreme events. Intensive residential and commercial development in such regions puts life and property at risk.

Key Issue: Threats to Estuarine Health

Estuaries are extremely productive ecosystems that are affected in numerous ways by climate. Winter temperatures are projected to continue to increase more than summer temperatures, resulting in a narrowing of the annual water temperature range of many estuaries. This is likely to cause species' ranges to shift and increase the vulnerability of some estuaries to non-native invasive species. Either increases or decreases in runoff would very likely create impacts to estuaries. Increased runoff would likely deliver increased amounts of nutrients such as nitrogen and phosphorous to estuaries, while simultaneously incre the stratification between freshwater runoff and marine waters. B ent additions and increased stratification would increase the pote blooms of algae that deplete the water of oxygen, increasing stres grasses, fish, shellfish, and other living things on the bottom of lak and oceans. Decreased runoff would likely reduce flushing, decre of estuarine nursery zones, and allow predators and pathogens of penetrate further into the estuary.

Winter temperatures are projected to continue to increase more than summer temperatures, resulting in a narrowing of the annual water temperature range of many estuaries. This is likely to cause species' ranges to shift and increase the vulnerability of some estuaries to non-native invasive species.

Ocean Circulation Conveyor Belt

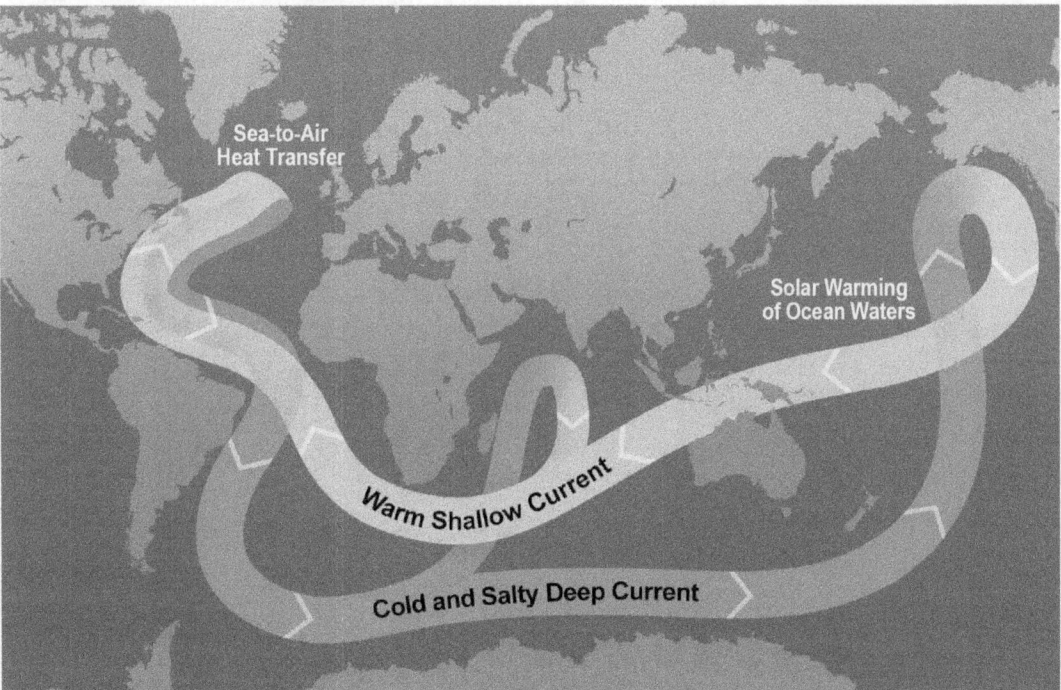

Sea-to-Air Heat Transfer

Solar Warming of Ocean Waters

Warm Shallow Current

Cold and Salty Deep Current

The ocean plays a major role in the distribution of the planet's heat through deep sea circulation. This simplified illustration shows this "conveyor belt" circulation which is driven by differences in heat and salinity. Records of past climate suggest that there is some chance that this circulation could be altered by the changes projected in many climate models, with impacts to climate throughout lands bordering the North Atlantic.

109

COASTAL AREAS AND MARINE RESOURCES KEY ISSUES

Coastal Wetland Survival

Coastal wetlands (marshes and mangroves) are highly productive ecosystems that are strongly linked to fisheries productivity. They provide important nursery and habitat functions to many commercially important fish and shellfish populations. Dramatic losses of coastal wetlands have already occurred on the Gulf Coast due to subsidence, changes caused by dams and levees that alter flow and reduce sediment supply, dredge and fill activities, and sea-level rise. Louisiana alone has been losing land at rates between 24 and 40 square miles per year during the last 40 years, accounting for as much as 80% of the total US coastal wetland loss.

In general, coastal wetlands will survive if soil buildup equals the rate of relative sea-level rise or if the wetland is able to migrate inland. However, if soil accumulation is unable to keep pace with high rates of sea-level rise, or if wetland migration is blocked by bluffs, coastal development, or shoreline protective structures (such as dikes, sea walls, and jetties), the wetland will be excessively inundated and eventually lost. The projected increase in the current rate of sea-level rise will very likely exacerbate coastal wetland losses nationwide, although the extent of impacts will vary among regions.

Processes Affecting Marsh Elevation

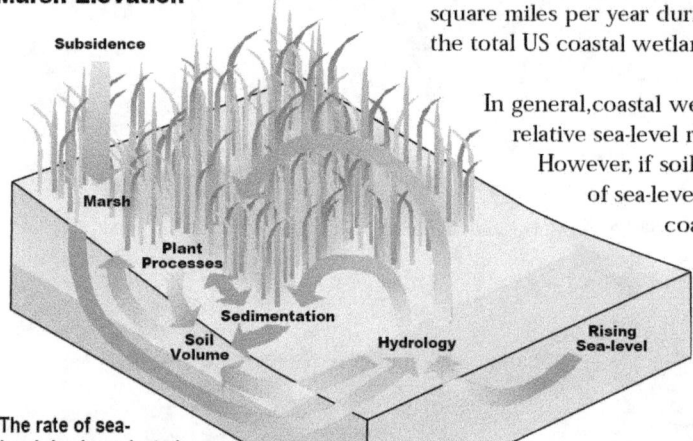

The rate of sea-level rise is projected to accelerate 2-5 fold over the next 100 years. The delivery of sediments to coastal wetlands is extremely important in determining the potential of these systems to maintain themselves in the face of current and future sea-level changes.

US Coastal Lands at Risk from a 20-inch Sea-Level Rise

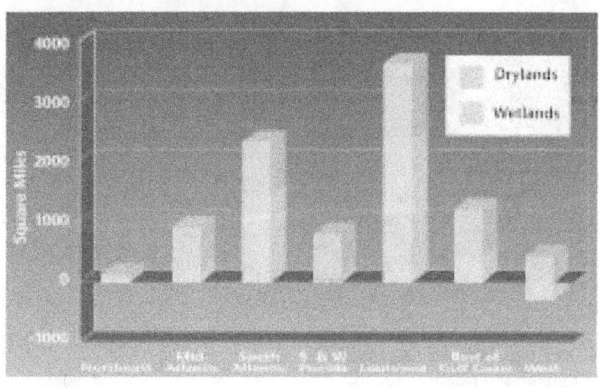

These bars show the square miles of coastal land at risk from a 20-inch rise in sea level, for seven areas of the US. Coastal wetlands projected to be inundated are shown in yellow while drylands projected to be inundated are shown in blue.

CO_2 and Temperature Stresses on Coral Reefs

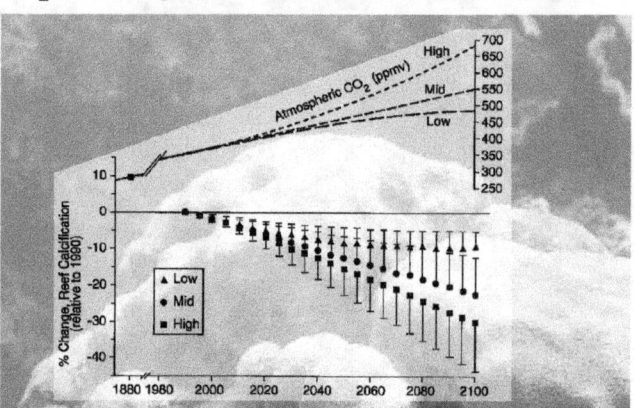

This graph shows a simulation of the effect of increased atmospheric CO_2 on percent change in coral reef calcification. Rising atmospheric CO_2 harms coral directly by making surface waters less alkaline, reducing corals' calcification and making their skeletons smaller and weaker. Warmer ocean temperatures will be another significant added stress, causing corals to expel the algae that live inside them and are crucial to their survival. Because these algae also give coral its color, this process is called "coral bleaching." Coral can recover after a short episode of warmer water, but if the warming persists the coral die. Under these combined stresses in addition to the existing stresses posed by human activities, corals may not survive in many areas.

Coral Reef Die-offs

Coral reefs play a major role in the environment and economies of two states (Florida and Hawaii) as well as most US territories in the Caribbean and Pacific. Coral reefs are valuable economic resources for fisheries, recreation, tourism, and coastal protection. In addition, reefs are one of the largest global storehouses of marine biodiversity, with untapped genetic resources. Some estimates of the global cost of losing coral reefs run in the hundreds of billions of dollars each year. The demise or continued deterioration of reefs could have profound implications for the US.

The last few years have seen unprecedented declines in the health of coral reefs. The 1998 El Niño was associated with record sea-surface temperatures and associated coral bleaching (when coral expel the algae that live within them and are necessary to their survival); in some regions, as much as 70% of the coral may have died in a single season. There has also been an upsurge in the variety, incidence, and virulence of coral diseases in recent years, with major die-offs in Florida and much of the Caribbean region. In addition, increasing atmospheric CO_2 concentrations could possibly decrease the calcification rates of the reef-building corals, resulting in weaker skeletons, reduced growth rates and increased vulnerability to erosion. Model results suggest that these effects would likely be most severe at the current margins of coral reef distribution.

There has also been an upsurge in the variety, incidence, and virulence of coral diseases in recent years, with major die-offs in Florida and much of the Caribbean region. In addition, increasing atmospheric CO_2 concentrations could possibly decrease the calcification rates of the reef-building corals, resulting in weaker skeletons, reduced growth rates and increased vulnerability to erosion.

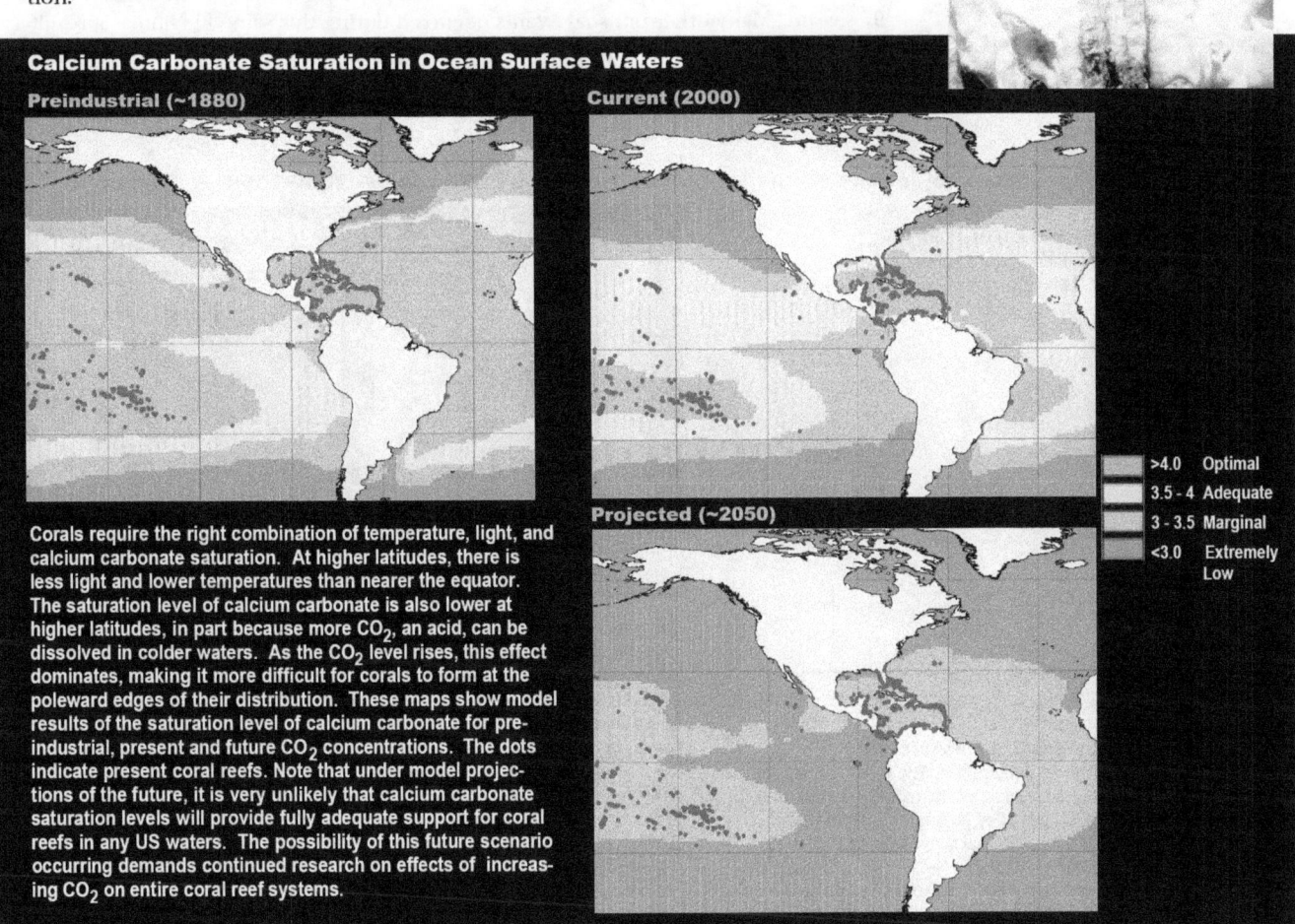

Calcium Carbonate Saturation in Ocean Surface Waters

Preindustrial (~1880)

Current (2000)

Projected (~2050)

>4.0	Optimal
3.5 - 4	Adequate
3 - 3.5	Marginal
<3.0	Extremely Low

Corals require the right combination of temperature, light, and calcium carbonate saturation. At higher latitudes, there is less light and lower temperatures than nearer the equator. The saturation level of calcium carbonate is also lower at higher latitudes, in part because more CO_2, an acid, can be dissolved in colder waters. As the CO_2 level rises, this effect dominates, making it more difficult for corals to form at the poleward edges of their distribution. These maps show model results of the saturation level of calcium carbonate for pre-industrial, present and future CO_2 concentrations. The dots indicate present coral reefs. Note that under model projections of the future, it is very unlikely that calcium carbonate saturation levels will provide fully adequate support for coral reefs in any US waters. The possibility of this future scenario occurring demands continued research on effects of increasing CO_2 on entire coral reef systems.

COASTAL AREAS AND MARINE RESOURCES KEY ISSUES

Stresses on Marine Fisheries

Atlantic and Gulf Coast shorelines are especially vulnerable to long term sea-level rise as well as any increase in the frequency of storm surges or hurricanes. Most erosion events on these coasts are the result of storms and extreme events, and the slope of these areas is so gentle that a small rise in sea level produces a large inland shift of the shoreline.

In the US,the total economic contribution of recreational and commercial fishing has been estimated at approximately $40 billion per year, with total marine landings averaging about 4.5 million metric tons over the last decade. Climate change is very likely to substantially alter the distribution and abundance of major fish stocks, and have important implications for marine populations and ecosystems.Changes over the long term are likely to include poleward shifts in distribution of marine populations.With changing ocean temperatures and conditions,shifts in the distribution of commercially important species are likely. For example,models suggest that several species of Pacific salmon are likely to have reduced distribution and productivity, while species that thrive in warmer waters,such as Pacific sardine and Atlantic menheden,are likely to have increased distribution.

Along the Pacific Coast,impacts to fisheries related to the El Niño/Southern Oscillation illustrate how climate directly affects marine fisheries on short time scales. For example,elevated sea-surface temperatures associated with the 1997-98 El Niño had a tremendous impact on the distribution and abundance of market squid,California's largest fishery by volume. Landings fell to less than 1,000 metric tons in the 1997-98 season,down from a record-breaking 110,000 tons in the 1996-97 season. Many other unusual events occurred during this same El Niño as a result of elevated sea-surface temperatures. Examples include widespread sea lion pup deaths in California,catches of warm-water marlin in the usually frigid waters off Washington State,and poor salmon returns in Bristol Bay, Alaska.

Sea-level Rise Projections

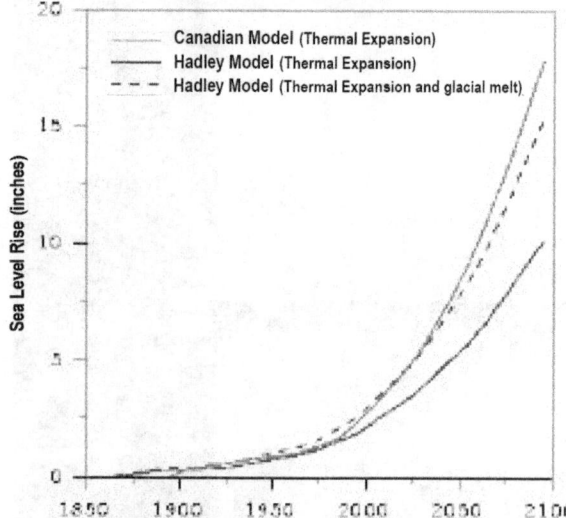

Historic and projected changes in sea level based on the Canadian and Hadley model simulations. The Canadian model projection includes only the effects of thermal expansion of warming ocean waters. The Hadley projection includes both thermal expansion and the additional sea-level rise projected due to melting of land-based glaciers. Neither model includes consideration of possible sea-level changes due to polar ice melting or accumulation of snow on Greenland and Antarctica.

Sea-level Rise Projections - 2080-2099

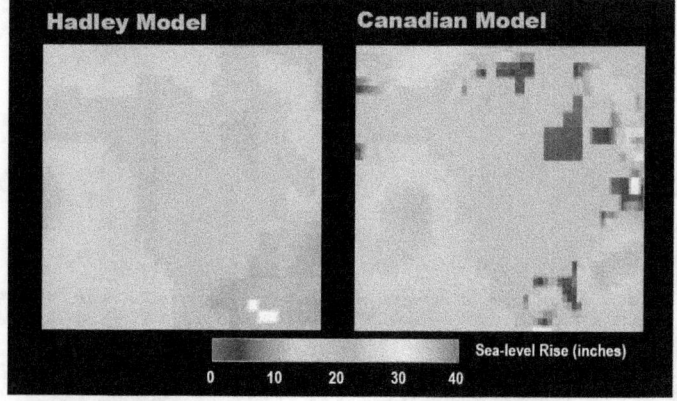

The Hadley model projects that ocean warming and melting of mountain glaciers will cause between 8 to 12 inches (20 to 30 cm) of sea-level rise by 2100 for much of the Atlantic and Gulf coasts of the US, depending on changes in winds and ocean current patterns. Projections for the Northeast US and the Pacific coast range from 13 to 16 inches (32 to 40 cm). Any effects of the rising or sinking of the coastal lands must be added to these numbers.

The Canadian model projects a more complex pattern of sea-level rise by 2100. Because of its larger warming estimate, sea level is projected to rise 20 to 24 inches (50 to 60 cm) along parts of the US Atlantic and Pacific coastlines. The orange peak in the Labrador Sea is the result of shifts in the location and intensity of ocean currents.

Adaptation Strategies

I t is difficult to assess the potential effects of climate change over the next few decades on coastal and marine resources,especially as climate variability is a dominant factor in shaping coastal and marine systems. The effects of future change will vary greatly in the diverse coastal regions of the nation. Additionally, human-induced disturbances also influence coastal and marine systems,often reducing the ability of systems to adapt,so that systems that might ordinarily be capable of responding to variability and change are less able to do so. In this context, climate change is likely to add to the cumulative impact of both natural and human-caused stresses on ecological systems and resources. This makes devising adaptation strategies particularly challenging.

With few exceptions,the potential consequences of climate change are not yet being considered in coastal management. It is especially urgent to begin adaptation now with regard to development of land in the coastal zone. In areas where beaches or wetlands must migrate inland to survive,it has been shown that implementing protection or retreat strategies for coastal developments can substantially reduce the economic impacts of inundation and shoreline movement. For example,coastal management programs in Maine,Rhode Island, South Carolina,and Massachusetts have implemented various forms of "rolling easement" policies to ensure that wetlands and beaches can migrate inland as sea level rises,and coastal landowners and conservation agencies can purchase the required easements. However, some regulatory programs continue to permit structures that may block the inland shift of wetlands and beaches. Additionally, allowing for such shoreline movement is only feasible in some locations due to the high degree of development on many coastlines.

With few exceptions, the potential consequences of climate change are not yet being considered in coastal management. It is especially urgent to begin adaptation now with regard to development of land in the coastal zone.

Annual Shoreline Change

This map is a preliminary classification of annual shoreline erosion throughout the US, in coarse detail and resolution. The areas most vulnerable to future sea-level change are those with low relief which are already experiencing rapid erosion rates, such as the Southeast and Gulf Coast.

■ Severely eroding   ▨ Moderately eroding   ☐ Relatively stable

FOREST SECTOR

F orests cover nearly one-third of the US, providing wildlife habitat, clean air and water, cultural and aesthetic values, carbon storage, recreational opportunities such as hiking, camping, fishing, and autumn leaf tours, and products that can be harvested such as timber, pulpwood, fuelwood, wild game, ferns, mushrooms, and berries. This wealth depends on forest biodiversity (the variety of plant and animal species) and forest functioning (water flows, nutrient cycling, and productivity). These aspects of forests are strongly influenced by climate. Native forests are adapted to their local climates; examples include the cold-tolerant boreal forests of Alaska, the summer-drought tolerant forests of the Pacific Northwest, and the drought-adapted piñon-juniper forests of the Southwest.

Human activities modify forests. Native forests have been converted to agricultural and urban uses. In some cases, forests have regrown on abandoned agricultural lands. Expansion of urban areas has fragmented forests into smaller, less-contiguous patches. Fire suppression has changed the species found in southeastern, midwestern, and western forests. Harvesting methods, where all trees or a few trees are cut, have also changed species composition. Trees have been planted for aesthetic and landscaping purposes in urban and rural areas that are often far outside of the species' natural range. Intensive management along with favorable climates in parts of the US has resulted in highly productive forests, such as southern pine plantations. Human activities will continue to modify forests while forests are also experiencing the effects of climate change.

Key Issue: Effects on Forest Productivity

S everal environmental factors that control the water and carbon balances of forests are changing rapidly and simultaneously. The global increases in atmospheric CO_2 concentrations are the best-documented factor. However, in some areas, other important atmospheric constituents are also increasing, including nitrogen oxides (a direct product of fossil fuel combustion that causes acid rain) and ground-level ozone ("smog," a product of chemical reactions between hydrocarbons and nitrogen oxides in the presence of sunlight).

A synthesis of laboratory and field studies and modeling indicates that forest productivity increases with the fertilizing effect of atmospheric CO_2 (see box in Agriculture section), but that these increases are strongly tempered by local conditions such as moisture stress and nutrient availability. Across a wide range of scenarios, it appears that modest warming could result in increased carbon storage in most forest ecosystems in the conterminous US. Yet under some warmer scenarios, forests, notably in the Southeast and the Northwest, could experience drought-induced losses of carbon, possibly exacerbated by increased fire disturbance. These potential gains and losses of carbon will be subject to changes in land-use, such as the conversion of forests to agricultural lands.

KEY ISSUES

- Effects on Forest Productivity

- Natural Disturbances such as Fire and Drought

- Biodiversity Changes

- Socioeconomic Impacts

Modest warming could result in increased carbon storage in most forest ecosystems in the conterminous US. Yet under some warmer scenarios, forests, notably in the Southeast and the Northwest, could experience drought-induced losses of carbon, possibly exacerbated by an increased fire disturbance.

Other components of environmental change, such as nitrogen deposition and ground-level ozone concentrations, also affect forest processes. Models identify a synergistic fertilization response between CO_2 and nitrogen enrichment, leading to further increases in productivity. Ozone, however, can suppress these gains. Current ozone levels, for example, have likely decreased production by 10% in Northeast forests and 5% in southern pine plantations. Interactions among these physical and chemical changes and other components of global change are important in determining the future of US forests.

Intensive management along with favorable climates in parts of the US has resulted in highly productive forests, such as the southern pine plantations. Human activities will continue to modify forests while forests are also experiencing the effects of climate change.

Current Distribution of Forests in the United States

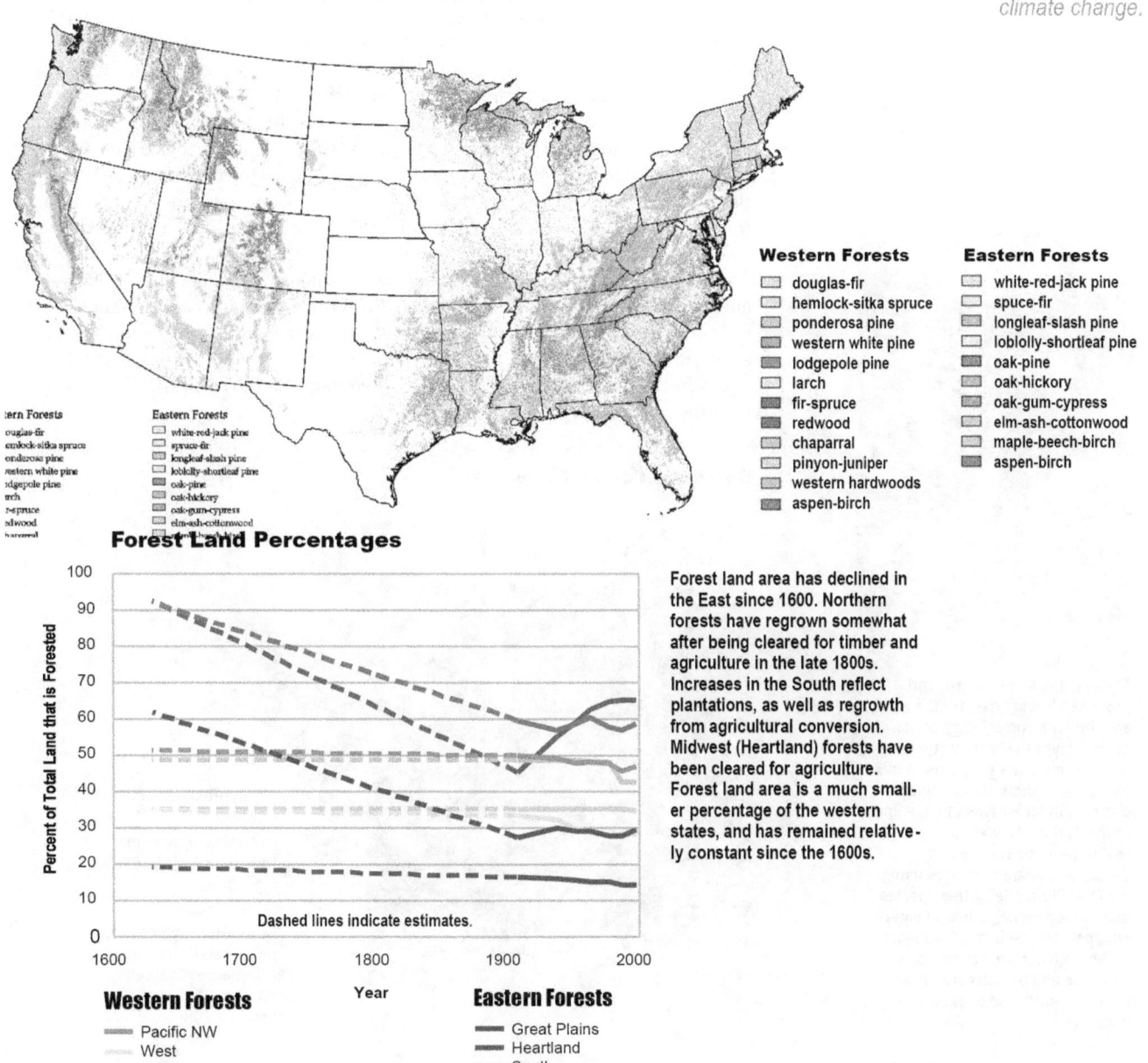

Western Forests
- douglas-fir
- hemlock-sitka spruce
- ponderosa pine
- western white pine
- lodgepole pine
- larch
- fir-spruce
- redwood
- chaparral
- pinyon-juniper
- western hardwoods
- aspen-birch

Eastern Forests
- white-red-jack pine
- spuce-fir
- longleaf-slash pine
- loblolly-shortleaf pine
- oak-pine
- oak-hickory
- oak-gum-cypress
- elm-ash-cottonwood
- maple-beech-birch
- aspen-birch

Forest Land Percentages

Forest land area has declined in the East since 1600. Northern forests have regrown somewhat after being cleared for timber and agriculture in the late 1800s. Increases in the South reflect plantations, as well as regrowth from agricultural conversion. Midwest (Heartland) forests have been cleared for agriculture. Forest land area is a much smaller percentage of the western states, and has remained relatively constant since the 1600s.

Dashed lines indicate estimates.

Y-axis: Percent of Total Land that is Forested (0–100)
X-axis: Year (1600–2000)

Western Forests
- Pacific NW
- West
- Alaska
- Islands

Year

Eastern Forests
- Great Plains
- Heartland
- South
- North

These maps show current and projected forest types for the eastern US. The current distribution of forest types reflects temperature and moisture gradients in this part of the nation. The simulated changes in forest types by the end of the 21st century are in response to the Hadley and Canadian climate scenarios using the DISTRIB model, a tree species distribution model. Pine-dominated types decline in the Southeast under both climate scenarios. Oak-pine and oak-hickory forest types are projected to expand northward.

FORESTS KEY ISSUES

Natural Disturbances Such as Fire and Drought

Natural disturbances having the greatest effects on forests include insects, disease, introduced species, fires, droughts, hurricanes, landslides, wind storms, and ice storms. Tree species have developed adaptations to some of these disturbances. For example, some tree species have developed very thick bark to protect them from repeated ground fires.

Over millennia, local, regional, and global changes in temperature and precipitation have influenced the occurrence, frequency, and intensity of these natural disturbances. These changes in disturbance regimes are a natural part of all ecosystems. However, forests may soon be facing rapid alterations in the nature of these disturbances as a consequence of climate change. For example, the seasonal severity of fire hazard is projected to increase about 10% over the next century over much of the US under both the Hadley and Canadian climate scenarios. Regionally, the Hadley scenario projects small decreases in fire hazard in the northern Great Plains, and the Canadian scenario projects a 30% increase in fire hazard for the southeastern US and Alaska.

The consequences of drought depend on annual and seasonal climate changes and whether the current drought adaptations offer resistance and resilience to new conditions. Under the Canadian and Hadley scenarios, the ecological models used in this Assessment indicate that increases in drought stresses will likely occur in the Southeast, southern Rocky Mountains and parts of the Northwest over the 21st century.

Dominant Forest Types

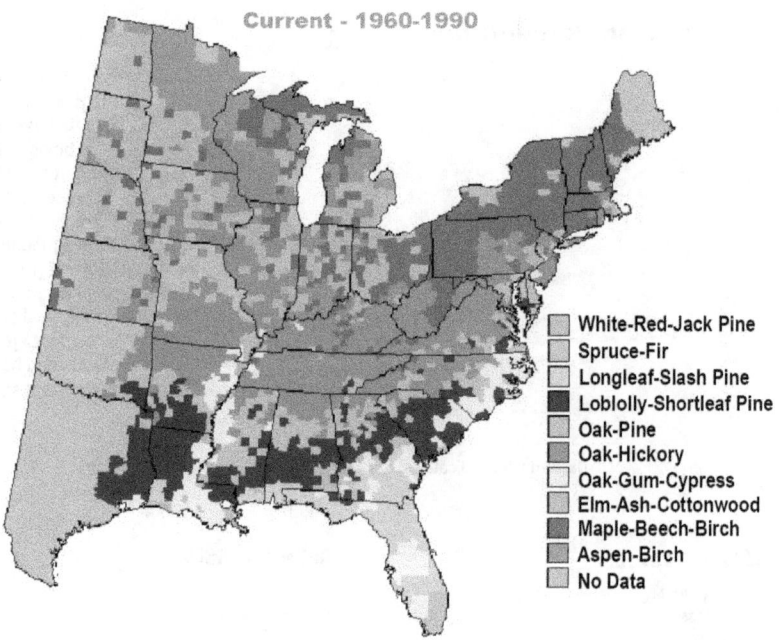

Current - 1960-1990

- White-Red-Jack Pine
- Spruce-Fir
- Longleaf-Slash Pine
- Loblolly-Shortleaf Pine
- Oak-Pine
- Oak-Hickory
- Oak-Gum-Cypress
- Elm-Ash-Cottonwood
- Maple-Beech-Birch
- Aspen-Birch
- No Data

The interactions between climate change and hurricanes,landslides,ice storms,wind storms,insects,disease, and introduced species are difficult to predict. But as climate changes, alterations in these disturbances and in their effects on forests are possible.

Biodiversity Changes

Changes in the distribution and abundance of plant and animal species reflect the birth, growth, death,and dispersal rates of individuals in a population. When aggregated, these processes result in the local disappearance or introduction of a species,and ultimately determine the species' range. While climate and soils exert strong controls on the establishment and growth of plant species,the response of plant and animal species to climate change will be the result of many interacting and interrelated processes operating over several scales of time and space. Migration rates, changes in disturbance regimes,and interactions within and between species will affect

the distribution of plants and animals. In addition,human activities influence the occurrence and abundance of species on the landscape.

Analyses of ecological models over several climate scenarios indicate that the location and area of the potential habitats for many tree species and communities are very likely to shift. Potential habitats for trees favored by cool environments are very likely to shift north. Habitats of alpine and sub-alpine spruce-fir could possibly be eliminated. Aspen,and eastern birch communities are likely to contract dramatically in the US and largely shift into Canada. Potential habitats that could possibly expand in the US are oak/hickory and oak/pine in the eastern US,and Ponderosa pine and arid woodland communities in the West.

How well these species track changes in their potential habitats will be strongly influenced by their dispersal abilities and the disturbances to these environments. Some native species will have difficulty dispersing to new habitats because of the rapid rate of

climate change and human land use along migration routes. For example, sagebrush and aspen communities are currently being reduced by conifer encroachment, grazing,invasive species,and urban expansion.

The effects of climate change on the rate and magnitude of disturbance (forest damage and destruction associated with fires,storms,droughts and pest outbreaks) will be an important factor in determining whether transitions from one forest type to another will be gradual or abrupt. If disturbances in New England, for example,do not increase,there is a possibility of a smooth transition from the present maple,beech,and birch tree species to oak and hickory. Where disturbances increase,transitions are very likely to be abrupt.

Invasive (weed) species that disperse rapidly are likely to find opportunities in newly forming communities. Thus,the species composition of these communities will likely differ substantially from those occupying similar habitats today.

Dominant Forest Types

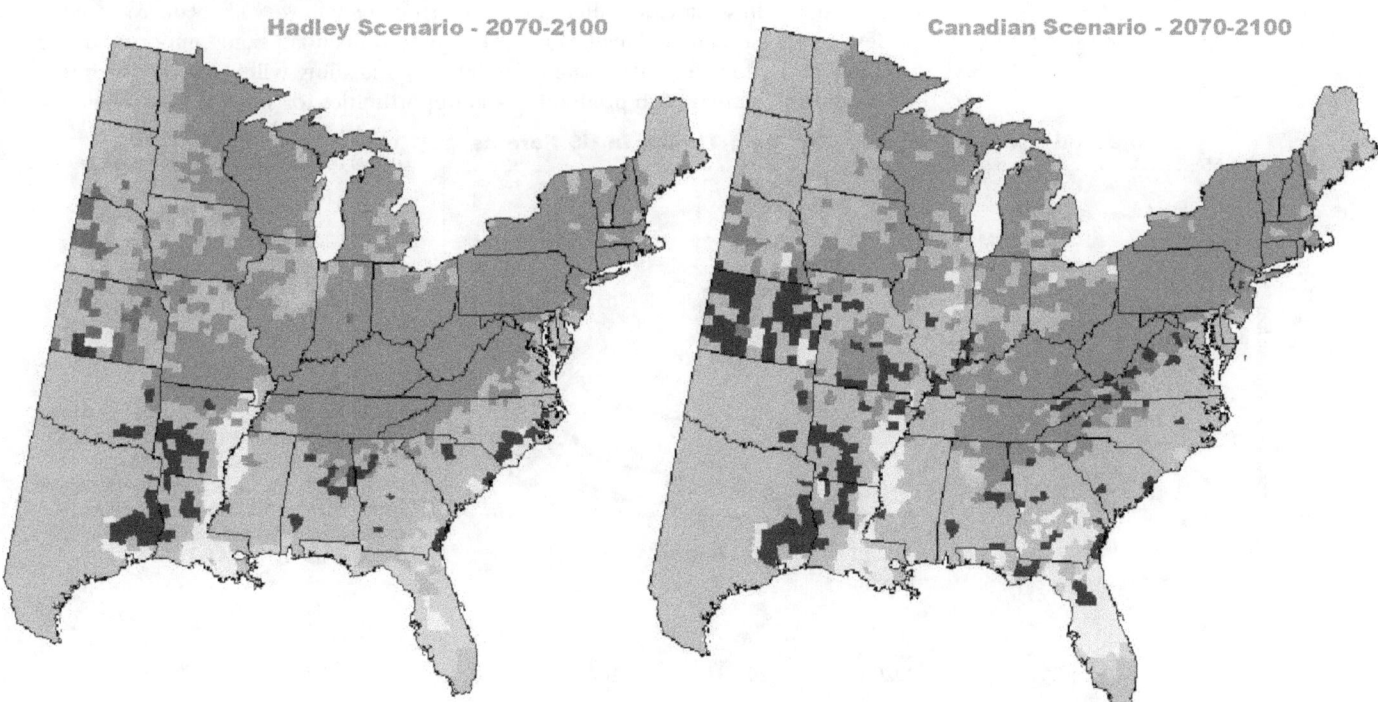

Hadley Scenario - 2070-2100

Canadian Scenario - 2070-2100

FORESTS KEY ISSUES

Socioeconomic Impacts

North America is the world's leading producer and consumer of wood products. The US has substantial exports of hardwood lumber, wood chips,logs, and some types of paper. However, the US also imports forest products,including 35% of its softwood lumber and more than half of its newsprint from Canada. The market for wood products in the US is highly dependent upon the future area in forests,species composition of forests,future supplies of wood,technological change in production and use, availability of substitutes such as steel and vinyl, demands for wood products,and competitiveness among major trading partners.

Analyses of the forest and agriculture sectors for a range of climate scenarios indicate that forest productivity gains are very likely to increase timber inventories over the next 100 years. Under these scenarios,the increased wood supply leads to reductions in log prices that,in turn,decrease producers'profits. At the same time, lower forest-product prices mean that consumers generally benefit. The projected net effect on the economic welfare of participants in both timber and agricultural markets increases about 1% above current values. Land will likely shift between forestry and agricultural uses as these economic sectors adjust to climate-induced changes in production. Although US total forest production generally increases, hardwood output is higher in all scenarios but softwood output increases only under moderate warming. Timber output increases more in the South than in the North.Sawtimber volume increases more than pulpwood volume.

It is very likely that outdoor recreation will be altered by climate change. Changes in benefits,as measured by aggregate days of activities and total economic value, will vary by type of recreation and location. In some areas,higher temperatures are likely to shift summer recreation activities,such as hiking,northward or to higher elevations. In winter,downhill skiing opportunities will very likely decrease with fewer cold days and reduced snowpack. Costs to maintain skiing opportunities are likely to rise in marginal climate areas. Effects on fishing will likely vary; warmer waters will increase fish production and opportunities for some warm water species,but decrease habitat and opportunities for cold water species.

Land will likely shift between forestry and agricultural uses as these economic sectors adjust to climate-induced changes in production.

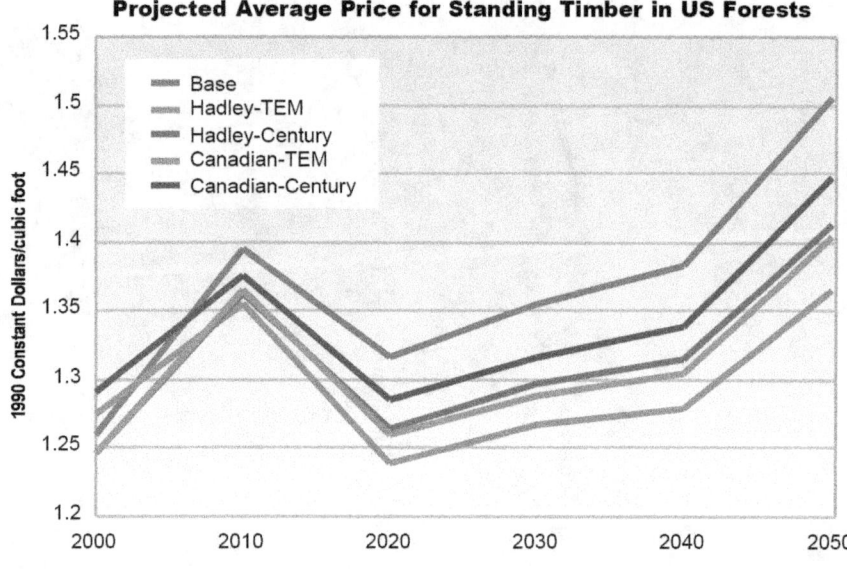

Projected Average Price for Standing Timber in US Forests

Legend:
- Base
- Hadley-TEM
- Hadley-Century
- Canadian-TEM
- Canadian-Century

Y-axis: 1990 Constant Dollars/cubic foot (1.2 to 1.55)
X-axis: 2000 to 2050

Prices for standing timber under all climate change scenarios remain lower than in a future without climate change (base). Prices under the Canadian scenario remain higher than prices under the Hadley scenario when either the TEM or the Century model is used.

Adaptation Strategies

Timber producers could possibly adjust and adapt to climate change under the scenarios used here, if new technologies and markets are recognized in a timely manner. Adaptations could include salvaging dead and dying timber and replanting species appropriate to a new climate.

While projected climate changes are likely to alter forests, the motivation for adaptation strategies will be strongly influenced by the level of economic activity in the US, population growth, tastes, and preferences including society's perceptions about these changes in forests. Market forces are powerful when it comes to land use and forestry, and as such, influence adaptation on private lands. However, for those forests valued for their current biodiversity, strategies to maintain these plant and animal species under climate change remain to be developed. It is possible that such strategies will be unavailable or impractical.

Markets for forest products adjust through altering prices for timber, wood, and paper products. The changes in climate and the consequent impact on forests will very likely change the market incentives for investment in intensive forest management (such as planting, thinning, genetic conservation, and tree improvement) and the incentive to develop and invest in wood-conserving technologies. Although these price changes are likely to alter consumption patterns (for example, substitution between wood and non-wood products), overall increase in the consumption of wood products very likely will still be predominantly influenced by population growth, the level of economic activity in the US and internationally, and personal preferences.

Timber producers could possibly adjust and adapt to climate change under the scenarios used here, if new technologies and markets are recognized in a timely manner. Adaptations could include salvaging dead and dying timber and replanting species appropriate to a new climate. The extent and pattern of timber harvesting and prices in the US will also be influenced by the global changes in forest productivity and prices of overseas products.

Potential climate-induced changes in forests must be put into the context of other human-induced pressures, which will undoubtedly change significantly over future decades. While the potential for rapid changes in natural disturbances could challenge current management strategies, these changes will co-occur with human activities such as agricultural and urban encroachment on forests, multiple use of forests, and air pollution.

Increased forest growth overall leads to increased wood supply; reductions in log prices decrease producers' welfare (profits), but generally benefit consumers through lower wood-product prices. Welfare is present value of consumer and producer surplus discounted at 4% for 2000-2100.

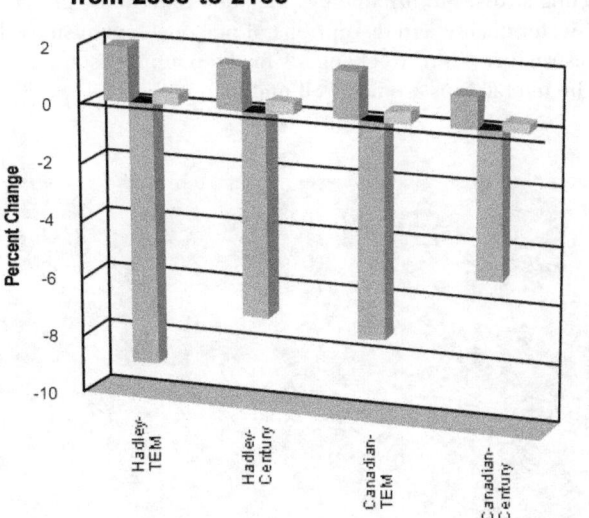

Change in Timber Product Welfare from 2000 to 2100

Percent Change

Hadley-TEM · Hadley-Century · Canadian-TEM · Canadian-Century

■ Timber product consumers
■ Timber product producers
■ Timber total welfare

CONCLUSIONS

Positive impacts will possibly be associated with climate changes such as warmer winters in the North, more precipitation in the Southwest, and longer growing seasons in parts of the nation where agriculture and forestry are important.

The warmer conditions associated with longer growing seasons in many parts of the country are likely to lead to increased pest and disease problems for agriculture and forestry.

Large Impacts in Some Places

The impacts of climate change will be significant for Americans. The nature and intensity of impacts will depend on the location, activity, time period, and geographic scale considered. For the nation as a whole, direct economic impacts are likely to be modest. However, the range of both beneficial and harmful impacts grows wider as the focus shifts to smaller regions, individual communities, and specific activities or resources. For example, while wheat yields are likely to increase at the national level, yields in western Kansas, a key US breadbasket region, are projected to decrease substantially under the Canadian climate model scenario. For resources and activities that are not generally assigned an economic value (such as natural ecosystems), substantial disruptions are likely.

Multiple-stresses Context

While Americans are concerned about climate change and its impacts, they do not think about these issues in isolation. Rather they consider climate change impacts in the context of many other stresses, including land-use change, consumption of resources, fire, and air and water pollution. This finding has profound implications for the design of research programs and information systems at the national, regional, and local levels. A true partnership must be forged between the natural and social sciences to more adequately conduct assessments and seek solutions that address multiple stresses.

Urban Areas

Urban areas provide a good example of the need to address climate change impacts in the context of other stresses. Although large urban areas were not formally addressed as a sector, they did emerge as an issue in most regions. This is clearly important because a large fraction of the US population lives in urban areas, and an even larger fraction will live in them in the future. The compounding influence of future rises in temperature due to global warming, along with increases in temperature due to local urban heat island effects, makes cities more vulnerable to higher temperatures than would be expected due to global warming alone. Existing stresses in urban areas include crime, traffic congestion, compromised air and water quality, and disruptions of personal and business life due to decaying infrastructure. Climate change is likely to amplify some of these stresses, although all the interactions are not well understood.

Impact, Adaptation, and Vulnerability

As the Assessment teams considered the negative impacts of climate change for regions,sectors,and other issues of concern,they also considered potential adaptation strategies. When considered together, negative impacts along with possible adaptations to these impacts define vulnerability. As a formula,this can be expressed as vulnerability equals negative impact minus adaptation. Thus,in cases where teams identified a negative impact of climate change, but could not identify adaptations that would reduce or neutralize the impact,vulnerability was considered to be high. A general sense emerged that American society would likely be able to adapt to most of the impacts of climate change on human systems but that the particular strategies and costs were not known.

Widespread Water Concerns

A prime example of the need for and importance of adaptive responses is in the area of water resources. Water is an issue in every region,but the nature of the vulnerabilities varies,with different nuances

in each. Drought is an important concern in every region. Snowpack changes are especially important in the West, Pacific Northwest,and Alaska. Reasons for the concerns about water include increased threats to personal safety, further reduction in potable water supplies,more frequent disruptions to transportation, greater damage to infrastructure,further degradation of animal habitat, and increased competition for water currently allocated to agriculture. The table below illustrates some of the key concerns related to water in each region.

Health, an Area of Uncertainty

Health outcomes in response to climate change are highly uncertain. Currently available information suggests that a range of health impacts is possible. At present, much of the US population is protected against adverse health outcomes associated with weather and/or climate, although certain demographic and geographic populations are at greater risk. Adaptation,primarily through the maintenance and improvement of public health systems and their responsiveness to changing climate conditions and to identified vulnera-

ble subpopulations should help to protect the US population from adverse health outcomes of projected climate change. The costs,benefits, and availability of resources for such adaptation need to be considered, and further research into key knowledge gaps on the relationships between climate/weather and health is needed.

Vulnerable Ecosystems

Many US ecosystems,including wetlands, forests, grasslands, rivers,and lakes, face possibly disruptive climate changes. Of everything examined in this Assessment,ecosystems appear to be the most vulnerable to the projected rate and magnitude of climate change,in part because the available adaptation options are very limited. This is important because,in addition to their inherent value,they also supply Americans with vital goods and services,including food, wood,air and water purification,and protection of coastal lands. Ecosystems around the nation are likely to be affected,from the forests of the Northeast to the coral reefs of the islands in the Caribbean and the Pacific.

WATER ISSUES	Region	Floods	Droughts	Snowpack/ Snowcover	Groundwater	Lake, river, and reservoir levels	Quality
	Northeast	X	X	X	X		X
	Southeast	X	X		X		X
	Midwest	X	X	X	X	X	X
	Great Plains	X	X	X	X	X	X
	West	X	X	X	X	X	X
	Northwest	X	X	X		X	
	Alaska		X	X			
	Islands	X	X		X		X

This table identifies some of the key regional concerns about water. Many of these issues were raised and discussed by stakeholders during regional workshops and other Assessment meetings held between 1997 and 2000.

CONCLUSIONS

Agriculture and Forestry Likely to Benefit in the Near Term

In agriculture and forestry, there are likely to be benefits due to climate change and rising CO_2 levels at the national scale and in the short term under the scenarios analyzed here. At the regional scale and in the longer term, there is much more uncertainty. It must be emphasized that the projected increases in agricultural and forest productivity depend on the particular climate scenarios and assumed CO_2 fertilization effects analyzed in this Assessment. If, for example, climate change resulted in hotter and drier conditions than projected by these scenarios, both agricultural and forest productivity could possibly decline.

Potential for Surprises

Some of the greatest concerns emerge not from the most likely future outcomes but rather from possible "surprises." Due to the complexity of Earth systems, it is possible that climate change will evolve quite differently from what we expect. Abrupt or unexpected changes pose great challenges to our ability to adapt and can thus increase our vulnerability to significant impacts.

A Vision for the Future

Much more information is needed about all of these issues in order to determine appropriate national and local response strategies. The regional and national discussion on climate change that provided a foundation for this first Assessment should continue and be enhanced. This national discourse involved thousands of Americans: farmers, ranchers, engineers, scientists, business people, local government officials, and a wide variety of others. This unique level of stakeholder involvement has been essential to this process, and will be a vital aspect of its continuation. The value of such involvement includes helping scientists understand what information stakeholders want and need. In addition, the problem-solving abilities of stakeholders have been key to identifying potential adaptation strategies and will be important to analyzing such strategies in future phases of the assessment.

The next phase of the assessment should begin immediately and include additional issues of regional and national importance including urban areas, transportation, and energy. The process should be supported through a public-private partnership. Scenarios that explicitly include an international context should guide future assessments. An integrated approach that assesses climate impacts in the context of other stresses is also important. Finally, the next assessment should undertake a more complete analysis of adaptation. In the current Assessment, the adaptation analysis was done in a very preliminary way, and it did not consider feasibility, effectiveness, costs, and side effects. Future assessments should provide ongoing insights and information that can be of direct use to the American public in preparing for and adapting to climate change.

The following section offers suggestions about new approaches, new knowledge, and new capabilities that would improve future assessments and thus provide more effective guidance for responding to the challenges posed by climate change.

This national discourse involved thousands of Americans: farmers, ranchers, engineers, scientists, business people, local government officials, and a wide variety of others. This unique level of stakeholder involvement has been essential to this process, and will be a vital aspect of its continuation. The value of such involvement includes helping scientists understand what information stakeholders want and need.

ECOSYSTEMS VULNERABILITY

The table below gives a partial list of potential impacts for major ecosystem types in various regions of the US. While the impacts are often stated in terms of what is likely to happen to plant communities, it is important to recognize that plant-community changes will also affect animal habitat.

Ecosystem	Goods	Services
Forests	timber, fuelwood, food (such as honey, mushrooms, and fruits)	purify air and water, generate soil, absorb carbon, moderate weather extremes and impacts, and provide wildlife habitat and recreation
Freshwater Systems	drinking and irrigation water, fish, hydroelectricity	control water flow, dilute and carry away wastes, and provide wildlife habitat, transportation corridors, and recreation
Grasslands	livestock (food, game, hides, fiber), water, genetic resources	purify air and water, maintain biodiversity, and provide wildlife habitat, employment, aesthetic beauty, and recreation
Coastal Systems	fish, shellfish, salt, seaweeds, genetic resources	buffer coastlines from storm impacts, maintain biodiversity, dilute and treat wastes, and provide harbors and transportation routes, wildlife habitat, employment, beauty, and recreation
Agro-ecosystems	food, fiber, crop genetic resources	build soil organic matter, absorb carbon, provide employment, and provide habitat for birds, pollinators, and soil organisms

Ecosystem Type

Impacts	NE	SE	MW	GP	WE	PNW	AK	IS
Forests								
Changes in tree species composition and alteration of animal habitat	X	X	X		X	X	X	X
Displacement of forests by open woodlands and grasslands under a warmer climate in which soils are drier		X						
Grasslands								
Displacement of grasslands by open woodlands and forests under a wetter climate					X			
Increase in success of non-native invasive plant species				X	X	X		X
Tundra								
Loss of alpine meadows as their species are displaced by lower-elevation species	X				X	X	X	
Loss of northern tundra as trees migrate poleward							X	
Changes in plant community composition and alteration of animal habitat							X	
Semi-arid and Arid								
Increase in woody species and loss of desert species under wetter climate					X			
Freshwater								
Loss of prairie potholes with more frequent drought conditions				X				
Habitat changes in rivers and lakes as amount and timing of runoff changes and water temperatures rise	X	X	X	X	X	X		
Coastal & Marine								
Loss of coastal wetlands as sea level rises and coastal development prevents landward migration	X	X			X	X		X
Loss of barrier islands as sea-level rise prevents landward migration	X	X						
Changes in quantity and quality of freshwater delivery to estuaries and bays alter plant and animal habitat	X	X			X	X	X	X
Loss of coral reefs as water temperature increases		X						X
Changes in ice location and duration alter marine mammal habitat							X	

RESEARCH PATHWAYS

New Approaches, New Knowledge, and New Capabilities for Our Nation

The National Assessment has defined a new vision for climate-impacts research. This vision has at its core a focus on integrated regional analysis and a close partnership of natural and social scientists with local, regional, and national stakeholders. Integrated analysis refers to considering the full range of stresses that affect a resource or system, including climate change and variability, land use change, air and water pollution, and many other human and natural impacts. For example, in studying water quality in a particular place, the direct and indirect effects of urban development, agricultural runoff, industrial pollution, and climate change-induced increases in heavy precipitation events all would need to be considered, along with many other factors. Integrated analysis also refers to integrating across all the relevant spatial scales for an issue, and these may extend from local to regional to national and even to global, depending on the issue. In the example of the local water quality study, this would mean integrating the effects of large-scale weather patterns on precipitation, as well as pollution inputs on both large and small scales, some of which originate far from the study area. Such integration across both multiple stresses and multiple scales is needed to provide the type of comprehensive analysis that decision-makers seek.

It is vital to our national interest that we meet these research needs so that we can, with increasing certainty, address the critical question: How vulnerable or resilient are the nation's natural and human resources and systems to the changes in climate projected to occur over the decades ahead?

Guided by this vision, the first National Assessment has identified a range of regional and sector vulnerabilities to climate variability and change that call for the attention of the American people and their leaders. In identifying vulnerabilities, the authors of the Assessment took great care to evaluate the likelihood of various climate-related outcomes. The likelihood of a number of outcomes was considered to be high. However, the likelihood of some of the other important potential outcomes was more difficult to judge due to lack of appropriate methods, uncertainties in knowledge, or shortcomings in research infrastructure such as computer power.

As our nation considers its strategies for dealing with climate-related vulnerabilities, scientists must work to reduce uncertainties underlying the vulnerability estimates. To assure that efforts to reduce uncertainties are efficient and to the point, the authors of the Assessment have identified a short list of priority research steps that are outlined below. These steps are organized into three categories: New Approaches, New Knowledge, and New Capabilities. It is vital to our national interest that we meet these research needs so that we can, with increasing certainty, address the critical question: How vulnerable or resilient are the nation's natural and human resources and systems to the changes in climate projected to occur over the decades ahead? With the new vision of regional analysis and scientist-stakeholder partnerships developed in the National Assessment, we have a powerful approach to effectively address this complex question.

Expanding New Approaches

This first National Assessment experimented with new approaches for linking the emerging findings and capabilities of the scientific community with the real-time needs of stakeholders who manage resources, grow food, plan communities, sustain commerce, and ensure public welfare. Teams were established in various regions across the country to look at how climate variability and change would affect particular locations. Other teams were established on particular topics that focused on how climate variability and change would affect issues of national significance. These types of efforts need to be sustained and expanded. In doing so we can build on a number of existing federal programs.

Recommendation 1: Develop a More Integrated Approach to Examining Impacts and Vulnerabilities to Multiple Stresses

The key requirement is to develop a truly integrated and widely accessible approach at regional and national scales appropriate for the examination of regional and national problems associated with biologic, hydrologic, and socioeconomic systems. A number of the regional and sector studies supporting the National Assessment have made important progress in such efforts but substantial additional efforts are required.

Expand the national capability to develop integrated, regional approaches of assessing impacts and vulnerabilities.

The regional teams supporting the National Assessment have produced new and innovative partnerships among a wide variety of scientists and stakeholders. In the process, they have catalyzed new modes of research and have demonstrated the potential of an integrated approach to assessing the consequences of climate variability and change. If the nation is to have improved projections of the impacts of and vulnerabilities to multiple stresses, we must accelerate the process of integrating research capabilities across the spectrum of natural and social sciences with the needs of public and private decision-makers. The importance of multiple stresses on specific environments and the importance of linkages between physical, biological, chemical, and human systems, require enhanced capabilities for regional analysis. The key elements of this strategy must be to (a) integrate observations at a regional level; (b) develop a comprehensive system designed to make the enormous amounts of data and information more accessible and useful to the public; (c) enable field and experimental studies that focus on solving regional problems; and (d) develop a foundation for building tractable, high-resolution coupled models that can address the outcomes associated with multiple stresses unique to each area of the country. Regional assessments add impetus for developing a comprehensive integrated approach, and this integration will engender substantial new capabilities to address the relationships between climate and air quality, energy demand, water quality and quantity, species distribution, ecosystem character, ultraviolet radiation, and human health indices in specific regions.

An integrated research approach would address such questions as:

How will the combination of high concentrations of ground-level ozone, high heat stress, and other factors affect human and plant health, especially in sensitive areas, such as the Great Smoky Mountains National Park, and metropolitan Houston?

RESEARCH PATHWAYS

Assessment of US impacts, including potential benefits, will increasingly require an examination of changes and response strategies around the world, and the manner in which these are translated to the global marketplace and environment.

Perform integrated national investigations of additional sectors and issues.

The choice of major sectors in this Assessment (water, forests, human health, coastal regions, and agriculture) reflected the fact that they were likely to be informative and important. Some of the themes were represented by a strong foundation of methods and models (e.g., agriculture) while others represented capabilities at very early stages of development (human health). These themes yielded a number of future research needs (e.g., the need to better understand human health relationships to extreme temperatures, other extreme weather events, and air quality, and to better characterize the relationships between climate and disease vectors and then between vector distribution and disease). In addition, a large number of important sectors and themes were not addressed, including climate impacts on transportation, energy, urban areas, and wildlife. These will require future investment in supporting research and assessment.

Consider international linkages in assessing national impacts.

In some cases, the vulnerability and resiliency of the US to climate impacts are highly dependent on the nature of the changes in other countries. For example, in the case of agriculture, the nature of the impacts strongly depends on international markets, and therefore the production and distribution of major crops around the world. These markets will reflect the extent of temperature and precipitation changes in other nations and the ability of these nations to cope with climate change and variability. Assessment of US impacts, including potential benefits, will increasingly require an examination of changes and response strategies around the world, and the manner in which these are translated to the global marketplace and environment.

Recommendation 2: Develop New Ways to Assess the Significance of Global Change to People

New methods for examining the potential impacts of climate change, adaptation options, and the vulnerability of communities, institutions, and sectors are essential to improving the assessment process. Research on these issues would result in a far greater ability to anticipate possible surprises, incorporate socioeconomic data in our analyses, and provide information that is useful for public and private decision-makers. The key research requirements involve improving methods to:

Understand and assign value to large and non-market impacts (e.g., on communities, resources, and ecosystems).

Changes that occur in natural and managed ecosystems, natural resources, and the other sectors are important because people assign them value, either in market or non-market terms. It is crucial to develop new ways to assign values to possible future changes in resources and ecosystems, especially in the cases of very large impacts and of processes and services that do not produce marketable goods. A focus on large impacts and non-market systems should provide insights that would enable decision-makers to understand the potential consequences of environmental change, as well as the potential consequences of particular adaptation or mitigation decisions.

Represent, analyze, and report uncertainties.

Describing scientific uncertainty is a task that faces every assessment of impacts or vulnerability. Findings in this Assessment,and their associated uncertainties,are based on the considered judgment of the NAST and on the peer-reviewed literature. The assessment process should extend the capabilities of decision-makers to understand potential uncertainties. For this reason,a range of additional methods for representing,analyzing,and reporting uncertainties should become a research focus.

Assess potential thresholds and breakpoints.

Some ecosystems and human institutions do not respond to rapid changes or stresses in continuous ways;if the stress exceeds certain thresholds,the system changes very rapidly, and sometimes in irreversible ways. It is very important to understand these types of responses because they raise particularly difficult challenges for adaptation. Using climate scenarios to help determine the conditions under which such changes might occur is therefore extremely important to pursue,because it can provide information of direct utility to decision-makers.

Develop and apply internally consistent socioeconomic futures
for use in assessing impacts.

In order to consider adaptation responses and ultimate vulnerabilities to climate change and other environmental stresses,assessments need to consider alternative possible socioeconomic and climatic futures using scenarios,probability distributions,or other methods. This Assessment began this process,but new methods to develop and apply such futures will improve the quality of the evaluation of the potential for adaptation in sector, regional and national analyses.

Developing New Knowledge

Determining how climate change will affect us necessarily builds on a wide array of scientific knowledge,not just of how the atmosphere works,but of how land ecosystems,the oceans,society, and many other aspects of the Earth system interact. Building this base of knowledge was the reason for establishing the US Global Change Research Program,and for the many programs and projects that it supports. Findings of this research have been essential to the overall undertaking of the Assessment. However, in the course of this Assessment,a number of areas have been identified where specific types of new knowledge are needed to assist society in preparing for the changing conditions of the 21st century. Improving projections of the responses of ecosystems,societal and economic systems,and climate, would improve scientists' ability to answer questions that are important to decision-makers.

The recommended research could help identify key thresholds beyond which certain ecosystems would no longer perform services people rely upon. For example, at what point would a particular lake no longer provide habitat for certain types of fish due to changes in climate, hydrology, and excess nutrient input from fertilizer runoff?

RESEARCH PATHWAYS

The nature of the response of complex natural and managed ecosystems to multiple stresses is one of the most important challenges to providing more certain projections of the impacts of climate variability and future climate change. Scientific studies are needed to extend our knowledge of many types of interrelationships between climate and ecosystems. These complex, interdependent interactions determine how organisms will respond to climate and other stresses and determine the potential vulnerability and/or resilience of these systems. Areas requiring intensified research to address this challenge include:

Terrestrial and aquatic natural ecosystem responses to multiple stresses, including the consequences for productivity, biodiversity, and other ecosystem processes and services.

Information is lacking on local, watershed, and continental scales to evaluate the responses of terrestrial and aquatic ecosystems to combinations of environmental stresses. Experimentally and observationally based investigations are needed of combinations of important environmental changes (such as changes in: CO_2 concentration; climate variability, temperature, land use, air and water quality, and species composition) as they affect important ecological processes (including productivity and nutrient cycling), attributes (such as species diversity, the responses and interactions of important individual plant and animal species, and the interactions among plant and animal communities), and services (such as regulating runoff). Results from such field experiments are needed as inputs for more sophisticated generations of ecosystem models to provide the information needed to project the responses of ecosystems to combinations of stresses, rather than separately treating individual stresses.

Managed ecosystem responses to multiple stresses, including their consequences for water quality and runoff, soil fertility, agricultural and forest productivity, and pest, weed, crop, and pathogen interactions through the development of integrated observations, process studies, and models.

Effective management of agricultural lands, forested ecosystems, and watersheds is closely coupled to the influences of pests, pathogens, climate, and other environmental variables. These interactions have not been addressed adequately in an experimental fashion. Research is needed on the interactions of these factors as they affect crop and forest productivity, soil fertility, water quality and quantity, the spread of pathogens and weeds, etc. Results from such experiments would provide insight into potential vulnerabilities from the combinations of environmental stresses, and input to better management responses to changes in those stresses.

The importance and interactions of climate, land cover, and land use in nutrient cycling, water supply and quality, runoff, and soil fertility.

Current land use models include socioeconomic variables such as human population size, affluence, and culture. However, the influences of climate change on land use, and of land use changes on regional climate are not adequately considered.

The recommended ecosystem research could address such questions as:

How will climate change combine with increased demand for agricultural production and increased urbanization to affect the rate of species endangerment in the US?

Will climate change and the increase in atmospheric CO_2 stimulate carbon sequestration in US forest ecosystems? If so, how much carbon will be sequestered and which forests will be the most effective carbon sinks?

Available analyses of ecological change tend to make one of two simplifications: either that a region is covered by its potential natural vegetation (the vegetation that would exist in the absence of human activities) or that the land-cover and land-use will remain as it is now, independent of climatic or other stresses. Both of these perspectives are limiting because they simplify the feedbacks between land use, ecological systems, and climate. New models of land-use change that integrate actual land-cover and land use information with ecological and economic processes would provide a crucial context for examining the potential consequences of human land-use decisions on a wide variety of ecosystem goods and services.

Better observations and models of ecosystem disturbance, and species dispersal and recruitment.

The ability to project changes in the ranges of important tree and plant species, and therefore changes in the make-up of forests and other ecosystems, is critically limited by information about the frequency of fires and other disturbances, the ability of seeds to disperse across current landscapes, and the factors that determine the success of plants in establishing themselves in new habitats and locations. Field observations, experimental studies, and historical analyses of past ecosystem changes are needed in order to understand both what is possible and the distances and rates of range changes that important species might actually achieve as climate changes. Such results would help fill crucial gaps in knowledge and enhance our ability to project the future ranges and distribution of important species.

Recommendation 4: Enhance Knowledge of How Societal and Economic Systems Will Respond to a Changing Climate and Environment

A greater understanding of the vulnerability and resilience of societal and economic systems is essential to addressing key uncertainties. Human roles in and responses to climate change and other environmental stresses are among the most important features of impact assessments. To assess real impacts on people and their societies, we should improve our understanding of how people and institutions will adapt to change and of the factors that will determine their vulnerability. Gaining such knowledge will require investing in the following key areas:

Understanding the resilience of communities, institutions, regions, and sectors (e.g., human health, urban areas, transportation, and international linkages).

The ability of communities, institutions, regions, and sectors to adapt has only begun to be addressed in this Assessment. Understanding how the capacity to adapt to a changing climate might be exercised, and therefore what vulnerabilities to climate change and other environmental stresses might remain, is an important next step in the human dimensions research agenda. The results of this research would enable a more integrated evaluation of both natural and social science aspects of human responses.

To assess real impacts on people and their societies, we need to improve our understanding of how people and institutions will adapt to change and of the factors that will determine their vulnerability.

RESEARCH PATHWAYS

The recommended research could help address questions such as:

What are the likely costs of adapting to increases in average temperature and heat index?

To what degree can cities take climate change into account in planning for new infrastructure, such as water distribution and routing, bridges, and peak power demands?

Improving understanding of how people and institutions have adapted to past climate variability and extreme events.

There is a wealth of information available on how people and institutions have responded to climate variability and other environmental changes in the past. New research that documents these responses, analyzes the underlying reasons for them, and explains how individual and institutional decisions were actually made will provide important insights into the feasibility of coping and adaptation options that might be available and considered in the future.

Greater information and analysis of specific potential adaptation options (e.g., costs, efficacy, time horizons, feasibility, and other impacts).

One of the critical unknowns in this Assessment's consideration of adaptation options stems from a lack of information about their potential costs, efficacy, time horizons required for implementation, other consequences, and feasibility. This type of information should be gathered as decision-makers consider specific adaptation options.

Recommendation 5: Refine our Ability to Project How Climate Will Change

This first National Assessment has revealed a number of key uncertainties in projecting climate change and variability at global, national, and regional scales. These uncertainties limit our ability to assess the responses of natural and managed ecosystems and societal and economic systems. Of greatest significance to decision-makers will be reducing uncertainties in several key areas by pursuing research that will lead to:

Improved understanding and analysis of the potential for future changes in severe weather, extreme events, and seasonal to interannual variability.

Many of the results in this Assessment demonstrate that changes in climate variability across a wide range of spatial and time scales have very important impacts on ecosystems, natural resources, and human systems. Long-term climate variations are strongly affected by how the oceans store and transport heat from warm regions near the equator to cold regions at high latitudes. For example, the El Niño-Southern Oscillation (ENSO) influences extreme weather and climate events by affecting the paths, frequency, and severity of winter and tropical storms. While model projections of ENSO and other sources of variability have advanced greatly over the past few years, much additional work is needed on how human-caused climate change might affect these patterns of variability. Much greater understanding is also needed about how climate change will influence the frequency, intensity, and likely locations of severe weather and climate events such as droughts, hurricanes, tornadoes, severe thunderstorms, and meteorological events that produce severe flooding.

Improved understanding of the spatial and temporal character of hydrologic processes, including precipitation, soil moisture, and runoff.

Some of the most important differences among climate model simulations involve projections of precipitation, soil moisture, and runoff, and these are of the greatest significance for ecosystems, agriculture, and water quality and quantity. Despite their importance, there is incomplete understanding of the physical processes that govern the water cycle and the extent to which these processes will be modified by climate change. The differences are sufficiently large in some regions of the country that we cannot even project whether there will be an increase or a decrease in soil moisture and runoff in these regions. A critical element of climate research must be an improved ability to simulate all aspects of the water cycle. Results of this research would substantially improve the estimates of potential vulnerabilities to climate change and other stresses.

Increased information on the nature of past climate, including its spatial and temporal character.

Model-generated scenarios of climate change and variability are only one way to examine potential futures. Another important method is to reconstruct the regional record of past changes in climate and their consequences in order to improve our understanding of how the natural world has operated in the past. These records illustrate the nature of past variability, provide an opportunity to assess climate model sensitivity, and offer insights into the response of ecological systems to past climate change. Results of this research would raise our confidence in the application of climatic information to evaluate impacts and vulnerability.

Adding New Capabilities

The nation needs a stronger capability for providing climate information that serves the national requirements for assessing vulnerabilities and impacts. A stronger national capability could deliver climate projections, including increased access to reliable model outputs and observational information, improved understanding of limitations, and greater availability of the specialized products required by increasingly sophisticated assessment science. Climate modeling and analysis are the foundation for developing climate scenarios that describe alternative futures for analysis of potential impacts of climate change, adaptation options, and vulnerability. Several steps can be taken in the near-term to enhance our capabilities to provide and use scenarios.

Recommendation 6: Extend Capabilities for Providing Climate Information

Addressing the broad spectrum of future societal needs will require continued improvements in observations, analysis, and the ability to forecast a wide variety of environmental variables. The elements required to develop comprehensive capability include:

Every three to seven years, the US is affected by a cycle of El Niño and La Niña events that leads to major changes in the frequency of hurricanes and in the paths of other weather systems that are responsible for extremes of precipitation and temperature across the US. The recommended research will enable future assessments to project how climate change will alter the frequency and severity of these extreme weather events.

RESEARCH PATHWAYS

A national modeling and analysis capability designed to provide long-term simulations, analysis of limitations and uncertainties, and specialized products for impact studies.

The nation's climate modeling expertise is widely recognized throughout the world and this expertise is dedicated to developing state-of-the-science model capability. Ensembles of long-term simulations, extending from the start of the robust historical record to at least the next 100 years, would provide important information to the nation. Further, the regional and sector teams in this Assessment have been requesting a host of specialized climate products that more directly tie future climate projections to specific decisions or vulnerabilities. The assessment process requires greater access to and a greater understanding of the limitations inherent in future projections in order to weigh the advantages and risks of alternative courses of action. Substantially higher model resolution is required to link climate with the scales of human decisions. The demand for these climate services exceeds the capabilities of the research functions of the nation's climate modeling centers.

Dedicated computer capability for developing ensemble climate scenarios, high-resolution models, and multiple emission scenarios for impact studies.

There is a need for ensemble climate simulations based on multiple-emission scenarios devoted to studies of climate impacts, vulnerabilities, and responses. The investment that is needed is to enhance the capacity of the climate modeling community to generate and analyze model runs that are dedicated for use by impact analysts. Similarly, future assessments need to investigate a range of plausible emissions and atmospheric concentrations of carbon dioxide and other greenhouse gases. The results of enhancing the capability to generate dedicated scenarios of emissions and climate would be a dramatic improvement in the range of outcomes that future assessments of vulnerability could analyze.

Reliable long-term observations and data archives.

One of the most often encountered limitations to the conduct of this Assessment, and one of the most often expressed needs of participants in the regional and sector assessment process, has been the lack of databases that truly reflect changes and variations in the environment, as opposed to those that unduly reflect uncertainties in observing methods. A commitment by the nation to provide integrated databases and information on multiple environmental conditions and trends, and indicators/measures of climate and related environmental changes, is essential to support and implement the research agenda. The US has tremendous potential to create more efficient and comprehensive measurement, archive, and data access systems that would provide greater scientific benefit to society by building upon existing weather and hydrologic stations and remote sensing capability, and integrating current efforts of local, state, and federal agencies. Improved data and information archives will substantially enhance future assessments.

A commitment to sustained high-quality observations is necessary for detecting changes in important aspects of our environment. New and improved data sets are required to address questions such as: how is our environment is being altered by climate change? And how much confidence can we place in future projections, given our ability to understand past changes and variations?

Addressing the Full Agenda

While the proposed research activities are all important individually, it will not be possible to substantially reduce our uncertainties and gaps in knowledge without consideration of their interconnections and interdependencies. The National Assessment Synthesis Team is convinced that the nation will benefit from multi-year investments in this focused program of research. Benefits will include major enhancement in our knowledge of the impacts of and vulnerabilities to global change on scales appropriate to the national interest and in our capacity to assess their importance.

We must accept the challenge to learn more and to conduct future assessments of multi-dimensional changes in climate, other environmental processes, and socioeconomic conditions. Meeting this challenge will require the new approaches, new knowledge, and new capabilities outlined above that can help reduce uncertainties while taking advantage of the great amount that we already know. Many of the building blocks of scientific knowledge, analytical capability, and commitment to the required integration are now in place. Through its regional, sector, and integrated approach, this Assessment has taken an important first step toward that future.

Areas with High Potential for Providing Needed Information in the Near-Term

Expand the national capability to develop integrated, regional approaches of assessing the impacts of multiple stresses, perhaps beginning with several case studies.

Develop capability to perform large-scale (over an acre) whole-ecosystem experiments that vary both CO_2 and climate.

Incorporate representations of actual land cover and land use into models of ecosystem responses.

Identify potential adaptation options and develop information about their costs, efficacy, side effects, practicality, and implementation.

Develop better ways to assign values to possible future changes in resources and ecosystems, especially for large changes and for processes and service that do not produce marketable goods.

Improve climate projections by providing dedicated computer capability for conducting ensemble climate simulations for multiple emission scenarios.

Focus additional attention on research and analysis of the potential for future changes in severe weather, extreme events, and seasonal to interannual variability.

Improve long-term data sets of the regional patterns and timing of past changes in climate across the US, and make these data-sets more accessible.

Develop a set of baseline indicators and measures of environmental conditions that can be used to track the effects of changes in climate.

Develop additional methods for representing, analyzing, and reporting scientific uncertainties related to global change.

We must accept the challenge to learn more and to conduct future assessments of multi-dimensional changes in climate, other environmental processes, and socioeconomic conditions. Meeting this challenge will require the new approaches, new knowledge, and new capabilities outlined above that can help reduce uncertainties while taking advantage of the great amount that we already know.

Appendix I - Biographical Sketches of NAST Members

Appendix I - Biographical Sketches of NAST Members

Jerry M. Melillo (co-chair)

Dr. Jerry M.Melillo (B.A. Wesleyan University, CT; Ph.D. Yale University) is in his twenty-fifth year as a research scientist at The Ecosystems Center of the Marine Biological Laboratory in Woods Hole, Massachusetts,and currently serves as the Center's Co-Director. Dr. Melillo's research on biogeo-chemistry includes work on global change,the ecological consequences of tropical deforestation, and sustainable management of forest ecosystems. He was a covening lead author on the 1990 and 1995 IPCC assessments of climate change. He has served as a vice-chair of the International Geosphere-Biosphere Programme (IGBP) and is currently President of ICSU's Scientific Committee on Problems of the Environment (SCOPE). Dr. Melillo founded the Marine Biological Laboratory's Semester in Environmental Science,an education program for undergraduates from small liberal arts colleges and universities in which students spend a term learning and doing environmental science in Woods Hole. Dr. Melillo also has a strong interest in science policy. He served as the Associate Director for Environment at the Office of Science and Technology Policy in the Executive Office of the President for 15 months in 1996 and 1997.

Anthony C. Janetos (co-chair)

Anthony C. Janetos is Sr. Vice President for Program at the World Resources Institute,an independ-ent policy research institute located in Washington,DC. He has held this position since 1999. He is an ecologist by training,with an A.B. in biology from Harvard,and M.A. and Ph.D. in biology from Princeton University. Dr. Janetos' expertise is in the interaction of ecosystems and atmospheric change. He has been a participant,lead author, and editor of many international scientific assess-ments,including chapters in IPCC Working Group I,the recent IPCC Special Report on Land-Use Change and Forestry, and the UNEP Global Biodiversity Assessment. He is the author or co-author of publications on the use of remote sensing to understand terrestrial ecosystems,the importance of biological diversity for ecosystem functioning,and the synergies among global environmental issues. Before coming to the World Resources Institute,Dr. Janetos managed research programs on the consequences of land-cover change for the National Aeronautics and Space Administration in its Office of Earth Science.

Thomas R. Karl (co-chair)

Thomas R.Karl is the Director of the National Oceanic and Atmospheric Administration's (NOAA) National Climatic Data Center within NOAA's National Environmental Satellite and Data Information Service (NESDIS). He also manages NOAA's Climate Change Data and Detection Program Element for NOAA's Office of Global Programs. He holds a Masters Degree in Meteorology from the University of Wisconsin. Mr. Karl is a fellow of the American Meteorological Society and the American Geophysical Union. He served as Chair (1997-1999) of the National Academy of Sciences Climate Research Committee. Mr. Karl has received numerous awards for his scholarly work on climate,including the Helmut Landsberg Award,the Climate Institute's Outstanding Scientific Achievements Award,and is a two-time recipient of the Department of Commerce's Gold Medal and recipient of their Bronze Medal,and the NOAA Administrator's Award. He currently is co-chair of NOAA's Decadal-to-Centennial Strategic Planning Team. He is also the Editor of the Journal of Climate and an Associate Editor of Climate Change. He has been a lead author on each of the IPCC's assessments of climate change since 1990. His special interests in areas of Earth Science Information include building homogenous data sets and providing stew-ardship for large data archives. Mr. Karl has authored nearly 100 peer-reviewed journal articles, been co-author or co-editor on numerous texts,and has published over 200 technical reports and atlases.

Eric J. Barron

Eric Barron is the Director of the Environment Institute in the Earth and Mineral Sciences college at Pennsylvania State University, where he is also Distinguished Professor of Geosciences .His areas of specialization include global change, numerical models of the climate system,and study of climate change throughout Earth history. He is also currently chair of the Board on Atmospheric Sciences and Climate of the National Research Council (NRC) as well as a member of the NRC committees on Global Change Research and Grand Challenges in the Environment.Dr. Barron received his bachelor's degree in Geology from Florida State University in 1973.He then began study of oceanography and climate at the Rosenstiel School of Marine and Atmospheric Sciences at the University of Miami, receiving his master's degree in 1976 and his Ph.D. in 1980.His career in climate modeling was initiated with a supercomputing fellowship at the National Center for Atmospheric Research (NCAR) in Boulder, Colorado in 1976.In 1980 he accepted a postdoctoral fellowship at NCAR,and in 1981 he joined the staff of the Climate Section.Dr. Barron returned to the University of Miami as an Associate Professor in 1985.In 1986 he became a member of the faculty of Pennsylvania State University, serving as Director of the Earth System Science Center and an Associate Professor of Geosciences.

Virginia Rose Burkett

Virginia Burkett is chief of the Forest Ecology Branch at the National Wetlands Research Center of the US Geological Survey (USGS),US Department of Interior, where she has worked since 1990. She also serves as an Associate Regional Chief Biologist for the USGS Central Region.Dr. Burkett supervises a team of wetland ecologists, forest scientists and landscape modelers who conduct research related to the ecology, management and restoration of forested wetlands.Her expertise includes wetland forest ecology and restoration,coastal wetland ecology, coastal management,and wildlife and fisheries management.Her current research involves bottomland hardwood regeneration in frequently flooded sites of the Mississippi River floodplain.Previously, Dr. Burkett served as director of the Louisiana Department of Wildlife and Fisheries,director of the Louisiana Coastal Zone Management program,and Assistant Director of the Louisiana Geological Survey. She received a B.S.in zoology and a M.S.in botany from Northwestern State University of Louisiana; her doctoral work in forestry was completed at Stephen F.Austin State University. Dr. Burkett is presently serving as a lead author of the chapter on global climate change and its impacts on coastal and marine ecosystems of the Third Assessment Report of the Intergovernmental Panel on Climate Change (IPCC).

Thomas F. Cecich

Thomas F. Cecich is Vice President of Environmental Safety for Glaxo Wellcome, where he has been employed for 15 years.In that capacity he is responsible for environmental protection and compliance,occupational safety and health,and emergency preparedness and response for this large multinational pharmaceutical company. Previously he held environmental and safety management positions at both the IBM and Allied Chemical corporations.Mr. Cecich has served as a faculty member in the Industrial Extension Service in the School of Engineering at North Carolina State University and an adjunct faculty member in the Department of Industrial Engineering.Mr. Cecich holds a B.S.in Industrial Engineering from the University of Miami and an M.S.in Industrial Engineering from North Carolina State University. He is certified in the practice of safety and industrial hygiene by the Board of Certified Safety Professionals and the American Board of Industrial Hygiene.He served on the Board of Certified Safety Professionals from 1993-1998 and was the President of the organization in 1997.He is the current Chairman of the Board of the Manufacturers and Chemical Industry Council of North Carolina,a state affiliate of the Chemical Manufacturers Association.

Appendix I - Biographical Sketches of NAST Members

Robert W. Corell (from January 2000)
Robert Corell is Senior Fellow at the Atmospheric Policy Program of the American Meteorological Society and Senior Research Fellow in the Belfer Center for Science and International Affairs of the Kennedy School of Government at Harvard University. Prior to these appointments in January 2000,he was Assistant Director for Geosciences at the National Science Foundation,where for over twelve years he had oversight for the Atmospheric,Earth,and Ocean Sciences and the Global Change programs of the National Science Foundation (NSF).While at the NSF, Dr. Corell also served as the Chair of the National Science and Technology Council's committee that has oversight of the US Global Change Research Program.Further, he served as chair and principal US delegate to many international bodies with interests in and responsibilities for climate and global change research programs.Dr. Corell is currently actively engaged in research concerned with both with the sciences of global change and with the interface between science and public policy. He currently serves as the Chair of the steering committee for the Arctic Climate Impact Assessment,which is an international assessment of the impacts of climate variability, change,and ultraviolet radiation increases in the Arctic region.Prior to joining the NSF in 1987,Dr. Corell was a Professor and academic administrator at the University of New Hampshire.Dr. Corell is an oceanographer and engineer by background and training,having received his Ph.D.,M.S.and B.S. degrees at the Case Institute of Technology and MIT and having held appointments at the Woods Hole Institution of Oceanography, the Scripps Institution of Oceanography, and the University of Washington.

Katharine L. Jacobs
Katharine Jacobs has been the Director of the Tucson Active Management Area of the Arizona Department of Water Resources since 1988.Her expertise is in groundwater management and developing practical,appropriate solutions to difficult public policy issues.She has worked in many capacities for the Department of Water Resources since 1981, verifying groundwater rights, developing mandatory conservation and enforcement programs,writing statewide rules requiring the use of renewable water supplies in new subdivisions,and working within the Tucson community building consensus solutions to serious water policy conflicts.She has facilitated development of groundwater recharge facilities and regional recharge policy. Ms. Jacobs has a bachelor's degree in biology from Middlebury College in Vermont,and a master's degree in environmental planning from the University of California,Berkeley. She participated in a National Research Council panel that authored the book Valuing Groundwater and has authored a number of publications on water management-related subjects.

Linda A. Joyce
Linda Joyce is Research Project Leader with the USDA Forest Service Rocky Mountain Research Station.She supervises a team of scientists who conduct research on the impact of terrestrial and atmospheric disturbances on alpine and forest ecosystems.She is also an affiliate faculty member in the Graduate Degree Program in Ecology and in the Rangeland Ecosystem Sciences Department,both programs at Colorado State University. Her research interests include modeling vegetation and ecosystem dynamics to assess the impact of climate change on ecosystem structure and function,quantifying the impacts of management on natural resources,linking ecological and economic analyses,and spatially optimizing natural resource production.Dr. Joyce serves as the Climate Change Specialist for the USDA Forest Service.She has contributed to the forestry and rangeland sections of the Intergovernmental Panel on Climate Change assessments.She received a bachelor's degree in mathematics from Grand Valley State University, a Master's in Environmental Science from Miami University of Ohio,and a Ph.D. in range ecology from Colorado State University

Barbara Miller

Dr. Miller is a Senior Water Resources Specialist in the World Bank's Africa Region, focusing on water resources management and international rivers in sub-Sahara Africa. She serves as a core member of the Nile Basin Initiative (NBI) Team, which is providing support to the ten countries that share the Nile River in the sustainable development and management of Nile water resources. Prior to joining the World Bank, Dr. Miller was President and Co-founder of Rankin International, a consulting firm providing engineering expertise in the areas of water, energy, the environment, and climate change. Previously, Dr. Miller spent ten years with the Tennessee Valley Authority (TVA). As Manager of the Flood Risk Reduction Department she was responsible for reducing flood damage potential in the Tennessee River Basin. While serving as a Senior Engineer at TVA's Engineering Laboratory, Dr. Miller managed the Reservoir System Analysis Group and was responsible for reservoir system modeling to support multipurpose reservoir system and hydropower operations. While at TVA, she also directed TVA's hydrologic modeling and climate change impact assessments. Dr. Miller has served on the advisory boards of several national climate change studies. Prior to TVA, Dr. Miller worked with the Illinois State Water Survey, Tippetts-Abbett-McCarthy-Stratton Engineers and Architects, and the US Environmental Protection Agency. Dr. Miller received her Ph.D. in Civil Engineering from the University of Illinois, Urbana-Champaign, and M.S. from the University of Wisconsin, Madison. She is a licensed Professional Engineer.

M. Granger Morgan

M. Granger Morgan is Head and Professor of Engineering and Public Policy at Carnegie Mellon University, where he also holds the Lord Chair Professor in Engineering as well as academic appointments in both the Department of Electrical and Computer Engineering and in The H. John Heinz III School of Public Policy and Management. He holds a B.A. from Harvard, where he concentrated in Physics, an M.S. in Astronomy and Space Science from Cornell, and Ph.D. from the Department of Applied Physics and Information Sciences at the University of California at San Diego. Much of Professor Morgan's recent research has focused on the integrated assessment of large complex policy problems that involve science and technology; in the treatment of uncertainty in quantitative policy analysis; in risk analysis, management and communication; and in several applied areas of technology and public policy. In collaboration with Hadi Dowlatabadi, he is active in Carnegie Mellon's Center for Integrated Study of the Human Dimensions of Global Change where his work includes better characterization of important uncertainties, and development of better policies to promote basic technology research in support of clean energy technology. Professor Morgan is a Fellow of the Institute of Electrical and Electronics Engineers, the American Association for the Advancement of Science, and the Society for Risk Analysis.

Appendix I - Biographical Sketches of NAST Members

Edward A. Parson (through January 2000)
Edward A. Parson is Associate Professor of Public Policy at Harvard's John F. Kennedy School of Government,where he has been on the Faculty since 1992.He is also a Faculty Research Associate in the Belfer Center for Science and International Affairs.Dr. Parson's research interests include the two related fields of environmental policy and negotiations.His environmental research stresses its international dimensions,including policy coordination,international institutions,negotiation,and conflict resolution.His book on the development of international cooperation to protect the ozone layer will appear in 2001.Current projects include work on scientific assessment in international policy-making;policy implications of carbon-cycle management;design of international market-based policy instruments;and development of policy exercises,simulation-gaming,and related novel methods for assessment and policy analysis.In negotiations, Parson's interests include the use of models and expert assessment bodies to support negotiations,learning and bargaining under uncertainty, and analysis of multi-party negotiations.He has developed a series of simulated multi-party negotiation exercises that are used for policy research and executive training in ten countries.Dr. Parson has served on the National Research Council Committee on Human Dimensions of Global Change.He holds degrees in Physics from the University of Toronto and Management Science from the University of British Columbia, and a Ph.D. in Public Policy from Harvard.He has worked and consulted for the White House Office of Science and Technology Policy, the US Congress Office of Technology Assessment,the US Environmental Protection Agency, Environment Canada,the Canadian Privy Council Office,the International Institute for Applied Systems Analysis,the Commission of the European Communities,and the UN Environment Programme.

Richard G. Richels
Richard Richels directs Global Climate Change Research at EPRI (formerly the Electric Power Research Institute) in Palo Alto,California. In previous assignments,he directed EPRI's energy analysis,environmental risk,and utility planning research activities.He has served on a number of national and international advisory panels,including committees of the Department of Energy, the Environmental Protection Agency, and the National Research Council.He has served as an expert witness at the Department of Energy's hearings on the National Energy Strategy and testified at Congressional hearings on priorities in global climate change research.He also served as a principal lead author for the Intergovernmental Panel on Climate Change's (IPCC) Second Assessment Report and is currently serving as a lead author for the IPCC's Third Assessment Report.He is a co-author with Alan Manne of Buying Greenhouse Insurance - the Economic Costs of CO_2 Emission Limits.Dr. Richels was awarded a M.S.degree in 1973 and Ph.D. degree in 1976 from Harvard University's Division of Applied Sciences.While at Harvard he was a member of the Energy and Environmental Policy Center.

David S. Schimel
David Schimel is Professor and Director in the Max-Planck-Institute for Biogeochemistry in Jena, Germany and a Senior Scientist at the National Center for Atmospheric Research in Boulder, Colorado.Previously, he has been a research scientist at Colorado State University in the College of Natural Resources and a National Research Council Senior Fellow at NASA-Ames Research Center. His scientific interests focus on the role of terrestrial ecosystems in the carbon cycle and on interactions between ecosystems and climate.He has served as Convening Lead Author of the IPCC for chapters on the Carbon Cycle and on impacts in North America.He has served on numerous committees and advisory panels including the National Research Council Committee on Global Change Research and interagency Carbon Cycle Science Plan Working Group.Dr. Schimel received a B.A.from Hampshire College in biology and applied mathematics,and Ph.D. from Colorado State University in Rangeland Ecosystem Science.

ACKNOWLEDGMENTS

About the National Science and Technology Council

President Clinton established the National Science and Technology Council (NSTC) by Executive Order on November 23,1993.This cabinet-level council is the principal means for the President to coordinate science,space,and technology policies across the Federal Government.The NSTC acts as a "virtual" agency for science and technology to coordinate the diverse parts of the Federal research and development enterprise. The NSTC is chaired by the President.Membership consists of the Vice President,the Assistant to the President for Science and Technology, Cabinet Secretaries and Agency Heads with significant science and technology responsibilities,and other senior White House officials.

An important objective of the NSTC is the establishment of clear national goals for Federal science and technology investments in areas ranging from information technology and health research,to improving transportation systems and strengthening fundamental research.The Council prepares research and development strategies that are coordinated across Federal agencies to form an investment package that is aimed at accomplishing multiple national goals. To obtain additional information regarding the NSTC,contact the NSTC Executive Secretariat at 202-456-6100 (voice).

About the Office of Science and Technology Policy

The Office of Science and Technology Policy (OSTP) was established by the National Science and Technology Policy, Organization,and Priorities Act of 1976.OSTP's responsibilities include advising the President on pol-

icy formulation and budget development on all questions in which science and technology are important elements;articulating the President's science and technology policies and programs;and fostering strong partnerships among Federal,State,and local governments,and the scientific communities in industry and academia.To obtain additional information regarding the OSTP, contact the OSTP Administrative Office at 202-456-6004 (voice).

The Committee on Environment and Natural Resources (CENR) is one of five committees under the NSTC.

D. James Baker, Co-Chair
National Oceanic and Atmospheric Administration

Rosina Bierbaum, Co-Chair
White House Office of Science and Technology Policy

Ghassem Asrar
National Aeronautics and Space Administration

James Decker
Department of Energy

Roland Droitsch
Department of Labor

Albert Eisenberg
Department of Transportation

Delores Etter
Department of Defense

Terrance Flannery
Central Intelligence Agency

George Frampton
Council on Environmental Quality

Kelley Brix
Department of Veteran's Affairs

Charles Groat
Department of the Interior

Len Hirsch
Smithsonian Institution

Kathryn Jackson
Tennessee Valley Authority

Eileen Kennedy
Department of Agriculture

Margaret Leinen
National Science Foundation

Paul Leonard
Housing and Urban Development

About the Committee on Environment and Natural Resources

The CENR is charged with improving coordination among Federal agencies involved in environmental and natural resources research and development,establishing a strong link between science and policy, and developing a Federal environment and natural resources research and development strategy that responds to national and international issues. To obtain additional information about the CENR,contact the CENR Executive Secretary at 202-482-5917 (voice).

Norine Noonan
Environmental Protection Agency

Kenneth Olden
Department of Health and Human Services

David Sandalow
Department of State

Wesley Warren
Office of Management and Budget

Craig Wingo
Federal Emergency Management Agency

Samuel Williamson
Office of the Federal Coordinator for Meteorology

SUBCOMMITTEES

Air Quality
Dan Albritton (NOAA), Chair
Bob Perciasepe (EPA), Vice Chair

Ecological Systems
Mary Clutter (NSF), Co-Chair
Don Scavia (NOAA), Co-Chair

Global Change
D. James Baker (NOAA), Chair
Ghassem Asrar (NASA), Vice Chair
Margaret Leinen (NSF), Vice Chair

Natural Disaster Reduction
Mike Armstrong (FEMA), Chair
John Filson (USGS), Vice Chair
Jaime Hawkins (NOAA), Vice Chair

Toxics and Risk
Norine Noonan (EPA), Chair
Bob Foster (DOD), Vice Chair
Kenneth Olden (HHS), Vice Chair

About the Subcommittee on Global Change Research

The Subcommittee on Global Change Research (SGCR) is one of five subcommittees under the Committee on Environment and Natural Resources (CENR). The SGCR is charged with improving coordination among Federal agencies participating in the U. S. Global Change Research Program (USGCRP), which was established by Congress in 1990 "to provide for development and coordination of a comprehensive and integrated United States research program which will assist the Nation and the world to understand, assess, predict, and respond to human-induced and natural processes of global change." The NAST is grateful for the SGCR establishing the NAST and providing oversight for its activities. To obtain additional information regarding the SGCR, contact the Office of the USGCRP at 202-488-8630 (voice) or see http://www.usgcrp.gov.

D. James Baker, Chair
(from January 2000)
National Oceanographic and
Atmospheric Administration
Department of Commerce

Robert W. Corell, Chair (through December, 1999)
National Science Foundation

Ghassem Asrar, Vice Chair
National Aeronautics and Space Administration

Margaret Leinen, Vice Chair
National Science Foundation

William Sommers
U. S. Forest Service
Department of Agriculture

Mary Gant
National Institute of Environmental Health Sciences, Department of Health and Human Services

Charles (Chip) Groat
Department of the Interior

J. Michael Hall
National Oceanic and Atmospheric Administration, Department of Commerce

Patrick Neale
Smithsonian Institution

Mark Mazur
Department of Energy

Margot Anderson
Department of Energy

Jeff Miotke
Department of State

Aristides A. Patrinos (SGCR liaison to the NAST)
Department of Energy

Fred Saalfeld
Department of Defense

Michael Slimak
Environmental Protection Agency

Executive Office Liaisons

Rosina Bierbaum
Office of Science and Technology Policy

Peter Backlund
Office of Science and Technology Policy

Steven Isakowitz
Office of Management and Budget

Sarah G. Horrigan
Office of Management and Budget

Ian Bowles
Council on Environmental Quality

The National Assessment Working Group is charged by the SGCR with overseeing and facilitating the coordination and preparation of national-scale assessments to document the current state of knowledge of the consequences of global change and their implications for policy and management decisions for the Nation. As such, they were the organizers and sponsors of the regional and sectoral assessments.

Paul Dresler
(through December 1999), Chair
Department of the Interior

Joel Scheraga, Vice Chair
Environmental Protection Agency

Richard Ball (through November 1999), Vice Chair
Department of Energy

Department of Agriculture
Margot Anderson (through March 2000)
Jeff Graham (through Sept. 1999)
Robert House
Fred Kaiser

Department of Defense and US Army Corps of Engineers
Thomas Nelson
Eugene Stakhiv

Department of Energy
Mitchell Baer (from November, 1999)
Jerry Elwood

Department of the Interior
Dave Kirtland
Ben Ramey

Environmental Protection Agency
Janet Gamble

National Aeronautics and Space Administration
Anne Carlson (from Jan. 2000)
Nancy Maynard (through Jan. 2000)
William Turner (from April 2000)
Louis Whitsett (through Dec. 2000)

National Institute of Environmental Health Sciences
Mary Gant

National Oceanic and Atmospheric Administration
Claudia Nierenberg
Roger Pulwarty
Caitlin Simpson

National Science Foundation
Thomas Spence

Office of Science and Technology Policy
Peter Backlund

ACKNOWLEDGMENTS

Lead Organizers of and Contributors to Regional and Sectoral Workshops and/or Assessments:

The information underpinning the National Assessment was generated through a series of workshops and assessments conducted by a distributed set of regional and sectoral teams. These teams were organized under the auspices of and sponsored by the USGCRP agencies. The sponsored workshops and assessments were conducted in an open manner and their findings were subjected to peer review by technical experts, interested stakeholders, other regional and sectoral teams, and the USGCRP agencies. Final responsibility for their findings rests with the individual teams. Information was channeled to the National Assessment Synthesis Team through a NAST liaison to the teams. Asterisks indicate the chair/co-chairs of the assessment teams. The NAST is particularly grateful for the extensive efforts of the regional and sectoral teams. Information on the regional and sectoral teams and their activities is available at http://www.nacc.usgcrp.gov. [Note: Sectoral workshops are not listed as many were held in conjunction with other meetings.]

REGIONAL WORKSHOP AND ASSESSMENT TEAMS

NORTHEAST MEGAREGION
Eric Barron, NAST liaison

Metropolitan East Coast
Workshop and Assessment Teams
Cynthia Rosenzweig*, National Aeronautics and Space Administration, Goddard Institute for Space Studies, and Columbia University
William Solecki*, Montclair State University
Carli Paine, Columbia University
Peter Eisenberger, Columbia University Earth Institute
Lewis Gilbert, Columbia University Earth Institute
Vivien Gornitz, Columbia University Center for Climate Systems Research
Ellen K. Hartig, Columbia University Center for Climate Systems Research

Douglas Hill, State University of New York, Stony Brook
Klaus Jacob, Lamont-Doherty Earth Observatory of Columbia University
Patrick Kinney, Columbia University Joseph A. Mailman School of Public Health
David Major, Columbia University Center for Climate Systems Research
Roberta Balstad Miller, Center for International Earth Science Information Network (CIESIN)
Rae Zimmerman, New York University Institute for Civil Infrastructure Systems, Wagner School

Mid-Atlantic
(Workshop September 9-11, 1997)
Ann Fisher*, Pennsylvania State University
David Abler, Pennsylvania State University
Eric J. Barron, Pennsylvania State University
Richard Bord, Pennsylvania State University
Robert Crane, Pennsylvania State University
David DeWalle, Pennsylvania State University
C. Gregory Knight, Pennsylvania State University
Ray Najjar, Pennsylvania State University
Egide Nizeyimana, Pennsylvania State University
Robert O'Connor, Pennsylvania State University
Adam Rose, Pennsylvania State University
James Shortle, Pennsylvania State University
Brent Yarnal, Pennsylvania State University

New England and Upstate New York
(Workshop September 3-5, 1997)
Barry Rock*, University of New Hampshire
Berrien Moore III*, University of New Hampshire
David Bartlett, University of New Hampshire
Paul Epstein, Harvard School of Public Health
Steve Hale, University of New Hampshire
George Hurtt, University of New Hampshire

Lloyd Irland, Irland Group, Maine
Barry Keim, New Hampshire State climatologist
Clara Kustra, University of New Hampshire
Greg Norris, Sylvatica Inc., Maine
Ben Sherman, University of New Hampshire
Shannon Spencer, University of New Hampshire
Hal Walker, EPA, Atlantic Ecology Division, Rhode Island

SOUTHEAST MEGAREGION
Virginia Burkett, NAST liaison

Central and Southern Appalachians
(Workshop May 26-29, 1998)
William T. Peterjohn (PI), West Virginia University
Richard Birdsey, USDA Forest Service
Amy Glasmeier, Pennsylvania State University
Steve McNulty, USDA Forest Service
Trina Karolchik Wafle, West Virginia University

Gulf Coast
(Workshop February 25-27, 1998)
Zhu Hua Ning*, Southern University and A & M College
Kamran Abdollahi*, Southern University and A & M College
Virginia Burkett, USGS National Wetlands Research Center
James Chamber, Louisiana State University
David Sailor, Tulane University
Jay Grymes, Southern Regional Climate Center
Paul Epstein, Harvard University
Michael Slimak, US Environmenatl Protection Agency

Southeast
(Workshop June 25-27, 1997)
Ron Ritschard*, University of Alabama – Huntsville
James Cruise*, University of Alabama – Huntsville
James O'Brien*, Florida State University
Robert Abt, North Carolina State University
Upton Hatch, Auburn University
Shrikant Jagtop, University of Florida
James Jones, University of Florida
Steve McNulty, USDA Forest Service

MIDWEST MEGAREGION
Tom Karl and David Easterling,
NAST liaisons

Eastern Midwest
(Workshop June 29-30,1998)
J. C.Randolph,Indiana University
Otto Doering,Purdue University
Mike Mazzocco,University of Illinois,
  Urbana - Champaign
Becky Snedegar, Indiana University

Great Lakes
(Workshop May 4-7,1998)
Peter J. Sousounis*,University of
  Michigan
Jeanne Bisanz*,University of Michigan
Gopal Alagarswamy, Michigan State
  University
George M.Albercook,University of
  Michigan
J. David Allan,University of Michigan
Jeffrey A.Andresen,Michigan State
  University
Raymond A.Assel,Great Lakes
  Environmental Research Laboratory
Arthur S.Brooks,University of
  Wisconsin-Milwaukee
Michael Barlage,University of Michigan
Daniel G.Brown,Michigan State
  University
H.H.Cheng,University of Minnesota
Anne H.Clites,Great Lakes
  Environmental Research Laboratory
Thomas E.Croley II,Great Lakes
  Environmental Research Laboratory
Margaret Davis,University of Minnesota
Anthony J. Eberhardt,Buffalo District,
  Army Corps of Engineers
Emily K.Grover, University of Michigan
Galina Guentchev, Michigan State
  University
Vilan Hung,University of Michigan
Kenneth E. Kunkel,Illinois State Water
  Survey
David A.R.Kirstovich,Illinois State
  Water Survey
John T. Lehman,University of Michigan
John D. Lindeberg,Center for
  Environmental Studies,Economics &
  Science
Brent M.Lofgren,Great Lakes
  Environmental Research Laboratory
James R.Nicholas,USGS,Lansing,
  Michigan
Jamie A.Picardy,Michigan State
  University
Jeff Price,American Bird Conservancy
Frank H.Quinn,Great Lakes
  Environmental Research Laboratory
Paul Richards,University of Michigan

Joe Ritchie,Michigan State University
Terry Root,University of Michigan
William B.Sea,University of Minnesota
David Stead,Center for Environmental
  Studies,Economics & Science
Shinya Sugita,University of Minnesota
Karen Walker, University of Minnesota
Eleanor A.Waller, Michigan State
  University
Nancy E.Westcott,Illinois State Water
  Survey
Mark Wilson,University of Michigan
Julie A.Winkler, Michigan State
  University
John Zastrow, University of Wisconsin

Additional Contributors
Stanley Changnon,Illinois State Water
  Survey

GREAT PLAINS MEGAREGION
Linda Joyce, NAST liaison

Central Great Plains
(Workshop May 27-29,1997)
Dennis Ojima*,Colorado State
  University
Jill Lackett*,Colorado State University
Dennis Child,Colorado State University
Alan Covich,Colorado State University
Celine Donofrio,Colorado State
  University
William Easterling, Pennsylvania State
  University
Kathy Galvin,Colorado State University
Luis Garcia,Colorado State University
Tom Hobbs,Colorado State
  University/State of Colorado Division
  of Wildlife
Martin Kleinschmit,Center for Rural
  Affairs
Kathleen Miller, National Center for
  Atmospheric Research
Jack Morgan,USDA Agricultural
  Research Service
Bill Parton,Colorado State University
Keith Paustian,Colorado State
  University
Gary Peterson,Colorado State
  University
Rob Ravenscroft, rancher, Nebraska
Lee Sommers,Colorado State University

Northern Great Plains
(Workshop November 5-7,1997)
George Seielstad*,University of North
  Dakota
Leigh Welling*,University of North
  Dakota
Kevin Dalsted,South Dakota State
  University

Jim Foreman,Ten Sleep, Wyoming
Robert Gough,Intertribal Council On
  Utility Policy
Janice Mattson,Precision Agriculture
  Research Association
James Rattling Leaf, Sinte Gleska
  University
Patricia McClurg,University of
  Wyoming
Gerald Nielsen,Montana State
  University
Gary Wagner, Climax,Minnesota
Pat Zimmerman,South Dakota School
  of Mines and Technology

Southern Great Plains
(Workshop May 24-25,1999)
Robert Harriss*,Texas A&M University
  (currently National Center for
  Atmospheric Research)
Tina Davies,Houston Advanced
  Research Center
David Hitchcock,Houston Advanced
  Research Center
Gerald North,Texas A&M University

Southwest-Rio Grande River Basin
  (Workshop March 2-4,1998)
Charles Groat,University of Texas-El
  Paso (currently US Geological
  Survey)
Honorable Silvestre Reyes,US House of
  Representatives,Texas

WEST MEGAREGION
Rich Richels, Barbara Miller and Joel
Smith, NAST liaisons

California
(Workshop March 9-11,1998)
Robert Wilkinson,* University of
  California,Santa Barbara
Jeff Dozier*,University of California,
  Santa Barbara
Richard Berk,University of California,
  Los Angeles
Dan Cayan,Scripps Institution of
  Oceanography, University of
  California,San Diego
Keith Clarke,University of California,
  Santa Barbara
Frank Davis,University of California,
  Santa Barbara
James Dehlsen,Dehlsen Associates
Peter H.Gleick, Pacific Institute for
  Studies in Development,
  Environment,and Security
Michael Goodchild,University of
  California,Santa Barbara
Nicholas Graham,Scripps Institution of
  Oceanography / University of

ACKNOWLEDGMENTS

California,San Diego

William J. Keese,California Energy Commission

Charles Kolstad,University of California,Santa Barbara

Michael MacCracken,USGCRP and Lawrence Livermore National Laboratory

Jim McWilliams,University of California,Los Angeles

John Melack,University of California, Santa Barbara

Norman L.Miller, Lawrence Berkeley National Laboratory / University of California,Berkeley

Harold A.Mooney, Stanford University

Peter Moyle,University of California, Davis

Walter C.Oechel,San Diego State University

Larry Papay, Bechtel Group

Claude Poncelet, Pacific Gas and Electric Company

Thomas Suchanek,NIGEC / University of California,Davis

Henry Vaux,University of California Office of the President

James R.Young,Southern California Edison

Rocky Mountain/Great Basin (Workshop February 16-18,1998)

Frederic Wagner*,Utah State University

Thomas Stohlgren*,US Geological Survey

Connely Baldwin,Utah State University

Jill Baron,US Geological Survey, Fort Collins,CO

Hope Bragg,Utah State University

Barbara Curti,Nevada Farm Bureau, Reno,NV

Martha Hahn, U.S.Bureau of Land Management,Boise,ID

Sherm Janke,Sierra Club,Bozeman,MT

Upmanu Lall,Utah State University

Linda Mearns,National Center for Atmospheric Research,Boulder, CO

Hardy Redd,Private Rancher, Lasal,UT

Gray Reynolds,Sinclair Corporation,Salt Lake City, UT

David Roberts,Utah State University

Lisa Schell,Colorado State University

Susan Selby, Las Vegas Valley Water District

Carol Simmons,Colorado State University

Dale Toweill,Idaho Dept.of Fish and Game,Boise,ID

Booth Wallentine,Utah Farm Bureau Federation,Salt Lake City, UT

Todd Wilkinson, Journalist,Bozeman, MT

Southwest-Colorado River Basin (Workshop September 3-5,1997)

William A.Sprigg*,University of Arizona

Todd Hinkley*,US Geological Survey

Diane Austin,University of Arizona

Roger C.Bales,University of Arizona

David Brookshire,University of New Mexico

Stephen P. Brown, Federal Reserve Bank of Dallas

Janie Chermak,University of New Mexico

Andrew Comrie,University of Arizona

Prabhu Dayal,Tucson Electric Power Company

Hallie Eakin,University of Arizona

David C.Goodrich,US Department of Agriculture

Howard P. Hanson,Los Alamos National Laboratory

Laura Huenneke,New Mexico State University

William Karsell, WAPA

Korine Kolivras,University of Arizona

Diana Liverman,University of Arizona

Rachel A.Loehman,Sandia National Laboratories

Jan Matusak,Metropolitan Water District of Southern California

Linda Mearns,National Center for Atmospheric Research

Robert Merideth,University of Arizona

Kathleen Miller, National Center for Atmospheric Research

David R.Minke,ASARCO

Barbara Morehouse,University of Arizona

Dan Muhs,US Geological Survey

Wilson Orr, Prescott College

Thomas Pagano,University of Arizona

Mark Patterson,University of Arizona

Kelly T. Redmond,Desert Research Institute

Paul R.Sheppard,University of Arizona

Verna Teller, Isleta Pueblo

James R.Young,Southern California Edison

NORTHWEST

Edward A. Parson, NAST liaison

(Workshop July 14-16,1997)

Philip Mote*,University of Washington

Douglas Canning,Department of Ecology, State of Washington

David Fluharty, University of Washington

Robert Francis,University of Washington

Jerry Franklin,University of Washington

Alan Hamlet,University of Washington

Blair Henry,The Northwest Council on Climate Change

Marc Hershman,University of Washington

Kristyn Gray Ideker, Ross and Associates

William Keeton,University of Washington

Dennis Lettenmaier, University of Washington

Ruby Leung, Pacific Northwest National Laboratory

Nathan Mantua,University of Washington

Edward Miles,University of Washington

Ben Noble,Battelle Memorial Institute

Hossein Parandvash, Portland Bureau of Water Works

David W. Peterson,US Geological Survey

Amy Snover, University of Washington

Sean Willard,University of Washington

ALASKA

Edward A. Parson, NAST liaison

(Workshops June 3-6,1997 and October 29-30,1998)

Gunter Weller*,University of Alaska Fairbanks

Patricia Anderson*,University of Alaska Fairbanks

Bronwen Wang*,US Geological Survey

Matthew Berman,University of Alaska Anchorage

Don Callaway, National Park Service

Henry Cole,Hydro Solutions & Purification LLC

Keith Criddle,Utah State University

Merritt Helfferich,Innovating Consulting Inc.

Glenn Juday, University of Alaska Fairbanks

Gunnar Knapp,University of Alaska Anchorage

Rosa Meehan, U. S. Fish and Wildlife Service

Thomas Osterkamp,University of Alaska Fairbanks

COASTAL AND ISLANDS MEGARE-
GION
Lynne Carter (from April 2000) and
Melissa Taylor (through March 2000),
NAST liaison

Pacific Islands
(Workshop March 3-6,1998)
Eileen Shea*,East-West Center
Michael Hamnett*,University of Hawaii
Cheryl Anderson,University of Hawaii
Anthony Barnston,NOAA,National
    Centers for Environmental Prediction
Joseph Blanco,Office of the Governor
    (State of Hawaii)
Kelvin Char, Office of the Governor
    (State of Hawaii) and NOAA National
    Marine Fisheries Service, Pacific
    Islands Area Office
Delores Clark,NOAA National Weather
    Service, Pacific Region Office
Scott Clawson,Hawaii Hurricane Relief
    Fund
Tony Costa, Pacific Ocean Producers
Margaret Cummisky, Office of the
    Honorable Daniel K.Inouye,United
    States Senate
Tom Giambelluca,University of Hawaii
Chip Guard,University of Guam
Richard Hagemeyer, NOAA National
    Weather Service, Pacific Region
    Office
Alan Hilton,NOAA Pacific ENSO
    Applications Center
David Kennard,FEMA Region IX, Pacific
    Area Office
Roger Lukas,University of Hawaii
Fred Mackenzie,University of Hawaii
Clyde Mark,Outrigger Hotels and
    Resorts-Hawaii
Gerald Meehl,National Center for
    Atmospheric Research
Jerry Norris, Pacific Basin Development
    Council
David Penn,University of Hawaii
Jeff Polovina,NOAA National Marine
    Fisheries Service
Roy Price,Hawaii State Civil Defense
    (retired)
Barry Raleigh,University of Hawaii
Kitty Simonds, Western Pacific Regional
    Fishery Management Council
Peter Vitousek,Stanford University
Diane Zachary, Maui Pacific Center

South Atlantic Coast and Caribbean
    (Workshop July 21-23,1998)
Ricardo Alvarez,International Hurricane
    Center
Krishnan Dandapani,Florida

International University
Shahid Hamid,Florida International
    University
Stephen Leatherman,International
    Hurricane Center
Richard Olson,International Hurricane
    Center
Walter Peacock,International Hurricane
    Center/Laboratory for Social and
    Behavioral Research
Paul Trimble,South Florida Water
    Management District

NATIVE PEOPLES/NATIVE
HOMELANDS
Michael MacCracken, NAST liaison

(Workshop October 28-November 1,
    1998)
Verna Teller, Isleta Pueblo
Robert Gough,Intertribal Council on
    Utility Policy
Schuyler Houser,American Indian
    Higher Education Consortium
Nancy Maynard, NASA
Fidel Moreno, Yaqui/Huichol
Lynn Mortensen,US Global Change
    Research Program
Patrick Spears,Lakota
Valerie Taliman,Navajo
Janice Whitney, HETF Fiduciary

Native Peoples/Native Homelands—
    Southwest
Stan Morain*,University of New Mexico
Rick Watson*,San Juan College
Diane Austin,University of Arizona
Mark Bauer, Diné College
Karl Benedict,University of New
    Mexico
Jennifer Bondick,University of New
    Mexico
Amy Budge,University of New Mexico
Linda Colon,University of New Mexico
Laura Gleasner, University of New
    Mexico
Jhon Goes In Center, Oglala Lakota
    Nation
Todd Hinckley, US Geological Survey
Doug Isely, Diné College
Bryan Marozas,DOI Bureau of Indian
    Affairs
Lynn Mortensen,US Global Change
    Research Program
Verna Teller, Isleta Pueblo
Carmelita Topaha,Navajo
Ray Williamson,George Washington
    University

Additional contributors
Patricia Anderson,University of Alaska
Lynne Carter, National Assessment
    Coordination Office
Schuyler Houser,American Indian
    Higher Education Consortium
Susan Marcus,US Geological Survey
Jeff Price,American Bird Conservancy
James Rattling Leaf, Sinte Gleska
    University
George Seielstad,University of North
    Dakota
Eileen Shea,East-West Center, Hawaii
Tony Socci,US Global Change Research
    Program
Leigh Welling,University of North
    Dakota

SECTOR
ASSESSMENT TEAMS

AGRICULTURE
Jerry Melillo, NAST liaison

John Reilly*,Massachusetts Institute of
    Technology
James Hrubovcak* (from October
    1999),US Department of Agriculture
Jeff Graham*,US Department of
    Agriculture (through Sept.1999)
David G.Abler, Pennsylvania State
    University
Robert Brown,Battelle-Pacific
    Northwest National Laboratory
Roy Darwin,US Department of
    Agriculture
Steven Hollinger, University of Illinois
Cesar Izaurralde,Battelle-Pacific
    Northwest National Laboratory
Shrikant Jagtap,University of Florida-
    Gainesville
James Jones,University of Florida-
    Gainesville
John Kimble,US Department of
    Agriculture
Bruce McCarl,Texas A&M University
Linda Mearns,National Center for
    Atmospheric Research
Dennis Ojima,Colorado State University
Eldor A. Paul,Michigan State University
Keith Paustian,Colorado State
    University
Susan Riha,Cornell University
Norman Rosenberg,Battelle-Pacific
    Northwest National Laboratory
Cynthia Rosenzweig, NASA-Goddard
    Institute for Space Studies
Francesco Tubiello, NASA-Goddard
    Institute for Space Studies

ACKNOWLEDGMENTS

COASTAL AREAS AND MARINE RESOURCES
Virginia Burkett, NAST liaison

Donald Boesch*,University of Maryland
Donald Scavia*,National Oceanic and Atmospheric Administration
John Field (project director),University of Washington
Robert Buddemeier, University of Kansas
Virginia Burkett, U. S.Geological Survey
Daniel Cayan,Scripps Institute of Oceanography
Michael Fogarty, University of Maryland
Mark A.Harwell,University of Miami
Robert Howarth,Cornell University
Curt Mason,National Oceanic and Atmospheric Administration
Richard A. Park,Eco-Modeling
Leonard J. Pietrafesa,North Carolina State University
Denise Reed,University of New Orleans
Thomas Royer, Old Dominion University
Asbury Sallenger, US Geological Survey
Michael Spranger, University of Washington
James Titus,Environmental Protection Agency

FORESTS
Linda Joyce, NAST liaison

Darius Adams,Oregon State University
John Aber*,University of New Hampshire
Steven McNulty*,US Department of Agriculture; Forest Service
Ralph Alig,US Department of Agriculture, Forest Service
Matthew P.Ayres,Dartmouth College
Dominique Bachelet,Oregon State University
Patrick Bartlein,University of Oregon
Carter J. Betz,US Department of Agriculture, Forest Service
Chi-Chung Chen,Texas A&M University
Rosamonde Cook,Colorado State University
David J. Currie,University of Ottawa, Canada
Virginia Dale,Oak Ridge National Laboratory
Raymond Drapek,Oregon State University
Michael D. Flannigan,Canadian Forest Service
Curt Flather, US Department of Agriculture, Forest Service
Andy Hansen,Montana State University
Paul J. Hanson,Oak Ridge National Laboratory
Mark Hutchins,Sno-Engineering,Inc
Louis Iverson,US Department of Agriculture, Forest Service
Lloyd Irland,The Irland Group
Linda Joyce,US Department of Agriculture, Forest Service
James Lenihan,Oregon State University
María Lombardero,Universidad de Santiago,Lugo,Spain
Ariel E.Lugo,US Department of Agriculture, Forest Service
Bruce McCarl,Texas A&M University
Ron Neilson,US Department of Agriculture, Forest Service
Chris J. Peterson,University of Georgia
Sarah Shafer, University of Oregon
Daniel Simberloff, University of Tennessee
Ken Skog,US Department of Agriculture, Forest Service
Brent L.Sohngen,Ohio State University
Brian J. Stocks,Canadian Forest Service
Frederick J. Swanson,US Department of Agriculture, Forest Service
Jake F.Weltzin,University of Tennessee
B.Michael Wotton,Canadian Forest Service

HUMAN HEALTH
Tom Cecich, NAST liaison

Michael A.McGeehin*,US Centers for Disease Control and Prevention
Jonathan A. Patz*, Johns Hopkins University School of Hygiene and Public Health
Susan M.Bernard (project director), Johns Hopkins University School of Hygiene and Public Health
Kristie L.Ebi,EPRI
Paul Epstein,Harvard Medical School
Anne Grambsch,US Environmental Protection Agency
Duane J. Gubler, US Centers for Disease Control and Prevention
Paul Reiter, US Centers for Disease Control and Prevention
Isabelle Romieu,US Centers for Disease Control and Prevention
Joan B.Rose,University of South Florida
Jonathan M.Samet, Johns Hopkins University School of Hygiene and Public Health
Juli Trtanj,National Oceanic and Atmospheric Administration

WATER RESOURCES
Katharine Jacobs, NAST liaison

D. Briane Adams*,US Geological Survey
Peter Gleick*, Pacific Institute for Studies in Development, Environment,and Security
Thomas O. Barnwell,US Environmental Protection Agency
Beth Chalecki, Pacific Institute for Studies in Development, Environment,and Security
Joseph Dellapenna, Villanova University
Ted Engman, NASA Goddard Space Flight Center
Kenneth D. Frederick,Resources for the Future
Aris P. Georgakakos,Georgia Institute of Technology
Donald R.Glaser,Water consultant
Gerald Hansler, Delaware River Basin Commission (retired)
Lauren Hay, US Geological Survey
Bruce P. Hayden,University of Virginia
Blair Henry,The Northwest Council on Climate Change
Steven Hostetler, US Geological Survey
Katharine Jacobs,Arizona Department of Water Resources
Sheldon Kamieniecki,University of Southern California
Debra S.Knopman,Center for Environmental Economics, Progressive Policy Foundation
Robert D. Kuzelka,University of Nebraska-Lincoln
Dennis Lettenmaier, University of Washington
Gregory McCabe,US Geological Survey
Judy Meyer, University of Georgia
Timothy Miller, US Geological Survey
Paul C."Chris" Milly, US Geological Survey, Geophysical Fluid Dynamics Laboratory
Norman Rosenberg,Battelle-Pacific Northwest National Laboratory
Michael J. Sale,Oak Ridge National Laboratory
Gregory E.Schwarz,US Geological Survey

John Schaake,National Oceanic and Atmospheric Administrration
Susan S.Seacrest,The Groundwater Foundation
Davis S.Shriner, US Forest Service
Eugene Z.Stakhiv, US Army Corps of Engineers
David M.Wolock,US Geological Survey

Scenario Development Teams:
As a basis for exploring the potential consequences of climate variability and change,information concerning climate,ecosystems,and socioeconomic factors was assembled to assist the regional and sectoral teams.As context for examining potential changes,information was assembled that documented conditions in the 20th century.To provide self-consistent estimates of how conditions might change in the future,simulations using state-of-the-art computer models were used to construct plausible scenarios of the types of conditions that might evolve during the 21st century.Asterisks indicate team leaders.Information on the various scenarios is available at http://www.nacc.usgcrp.gov.

CLIMATE VARIABILITY AND CHANGE
Eric Barron*, Pennsylvania State University
David Easterling*,NOAA National Climate Data Center
Benjamin Felzer*,National Center for Atmospheric Research
Tom Karl*,NOAA National Climate Data Center
Michael MacCracken*, USGCRP/National Assessment Coordination Office
Richard Ball,Department of Energy (retired)
Tony Barnston,NOAA,NCEP, Climate Prediction Center
Denise Blaha,University of New Hampshire
George Boer, Canadian Centre for Climate Modelling and Analysis, Victoria,BC
Ruth Carnell,Hadley Centre, Meteorological Office,Bracknell,UK
Aiguo Dai,National Center for Atmospheric Research
Christopher Daly, Oregon State University
Hank Fisher, National Center for Atmospheric Research
Greg Flato,Canadian Centre for Climate Modelling and Analysis,Victoria,BC

Byron Gleason,National Climatic Data Center
Jonathan Gregory, Hadley Centre, Meteorological Office,Bracknell,UK
Yuxiang He,NOAA,NCEP, Climate Prediction Center
Preston Heard,Indiana University - Bloomington
Roy Jenne,National Center for Atmospheric Research
Dennis Joseph,National Center for Atmospheric Research
Tim Kittel,National Center for Atmospheric Research
Richard Knight,NOAA National Climate Data Center
Steven Lambert,Canadian Centre for Climate Modelling and Analysis, Victoria,BC
Linda Mearns,National Center for Atmospheric Research
John Mitchell,Hadley Centre, Meteorological Office,Bracknell,UK
James Risbey, Carnegie Mellon University
Nan Rosenbloom,National Center for Atmospheric Research
J.Andy Royle,US Fish and Wildlife Service,Laurel MD
Annette Schloss,University of New Hampshire
Joel B.Smith,Stratus Consulting
Steve Smith, Pacific Northwest National Laboratory
Peter Sousounis,University of Michigan
David Viner, Climatic Research Unit, Norwich,UK
Warren Washington,National Center for Atmospheric Research
Tom Wigley, National Center for Atmospheric Research
Francis Zwiers,Canadian Centre for Climate Modelling and Analysis, Victoria,BC

ECOSYSTEMS
Timothy G. F. Kittel*,National Center for Atmospheric Research
Jerry Melillo*, Woods Hole Marine Biological Laboratory
David S.Schimel*,Max-Planck-Institute for Biogeochemistry, Jena,Germany
Steve Aulenbach,National Center for Atmospheric Research
Dominique Bachelet,Oregon State University
Sharon Cowling,Lund University, Sweden
Christopher Daly, Oregon State University
Ray Drapek,Oregon State University

Hank H. Fisher, National Center for Atmospheric Research
Melannie Hartman,Colorado State University
Kathy Hibbard,University of New Hampshire
Thomas Hickler, Lund University, Sweden
Cristina Kaufman,National Center for Atmospheric Research
Robin Kelly, Colorado State University
David Kicklighter, Marine Biological Laboratory
Jim Lenihan,Oregon State University
David McGuire, U.S.Geological Survey and University of Alaska, Fairbanks, AK
Ron Neilson,USDA Forest Service
Dennis S.Ojima,Colorado State University
Shufen Pan,Marine Biological Laboratory
William J.Parton,Colorado State University
Louis F.Pitelka,University of Maryland Appalachian Laboratory
Colin Prentice,Max-Planck-Institute for Biogeochemistry, Jena,Germany
Brian Rizzo,University of Virginia
Nan A.Rosenbloom,National Center for Atmospheric Research
J.Andy Royle, U. S.Department of the Interior
Steven W. Running,University of Montana
Stephen Sitch, Potsdam Institute for Climate Impact Research,Germany
Ben Smith,Lund University, Sweden
Thomas M.Smith,University of Virginia
Martin T. Sykes,Lund University, Sweden
Hanqin Tian,Marine Biological Laboratory
Justin Travis,Lund University, Sweden
Peter E.Thornton,University of Montana
F. Ian Woodward,University of Sheffield, UK

SOCIO-ECONOMIC CONDITIONS
Edward A. Parson*,Harvard University
Jae Edmonds, Pacific Northwest National Laboratory
Ann Fisher, Pennsylvania State University
Linda Joyce,US Forest Service, Department of Agriculture
Barbara Miller,World Bank
M.Granger Morgan,Carnegie Mellon University
Richard Richels,EPRI
Nestor Terlickij,NPA Data Associates

David Vogt,Oak Ridge National
Laboratory
Tom Wilbanks,Oak Ridge National
Laboratory
Sherry Wright,Oak Ridge National
Laboratory

Planning and Development Workshops and Activities:
To develop the plans for the assessment and ensure coordination among the various teams,a series of planning and development and other coordination activities were held.

ASPEN GLOBAL CHANGE INSTITUTE
(Aspen,CO, July 29 through August 7, 1997)
Michael MacCracken,Office of the US Global Change Research Program
William Easterling, Pennsylvania State University
Paul Dresler, Department of the Interior
John Katzenberger,Aspen Global Change Institute
Melissa Taylor, US Global Change Research Program

INTEREGIONAL FORUM
Tom Wilbanks,Oak Ridge National Laboratory, chair
Co-chairs from all of the regions
Lynne Carter, liaison from National Assessment Coordination Office
Paul Dresler, liaison from National Assessment Working Group
Joel Scheraga,liaison from National Assessment Working Group

STAKEHOLDER GUIDELINES
Tom Wilbanks,Oak Ridge National Laboratory and National Center for Environmental Decision-making Research (NCEDR)
David Cash,Harvard University
Nichole Kerchner, University of Tennessee
Robb Turner,Joint Institute for Energy and Environment
Amy Wolfe,Oak Ridge National Laboratory

US CLIMATE FORUM
(Washington DC,November 12-13, 1997)
Richard Ball,Department of Energy
Susan Bassow, Office of Science and Technology Policy
Rosina Bierbaum,Office of Science and Technology Policy

Robert Corell,National Science Foundation
Paul Dresler, Department of Interior
Ann Fisher, Pennsylvania State University
David Goodrich,National Academy of Sciences
Susan Gordon,Department of State
Blair Henry, Northwest Council on Climate Change
Michael MacCracken,National Assessment Coordination Office
Jerry Melillo,Marine Biological Laboratory
Wil Orr, City of Scottsdale,Arizona
Aristides Patrinos,Department of Energy
Joel Scheraga,Environmental Protection Agency
George Seielstad,University of North Dakota
Robert Shepard,Science and Engineering Alliance
Melissa Taylor, National Assessment Coordination Office

US NATIONAL ASSESSMENT 1998 WORKSHOP
(Monterey CA, July 27-31,1998)
Michael MacCracken,National Assessment Coordination Office
Melissa Taylor, National Assessment Coordination Office
Paul Dresler, Department of the Interior

US NATIONAL ASSESSMENT 1999 WORKSHOP
(Atlanta GA,April 12-15,1999)
Tom Wilbanks,Oak Ridge National Laboratory
Paul Dresler, Department of the Interior
Joel Scheraga,Environmental Protection Agency
Melissa Taylor, National Assessment Coordination Office

National Assessment Coordination Team:
To facilitate overall coordination among the NAST, the USGCRP agencies,and the regional and sectoral teams,the Subcommittee on Global Change Research established the National Assessment Coordination Office (NACO) and enlisted the assistance of others drawn from various organizations.The National Assessment Synthesis Team expresses its appreciation for their contributions.Information about NACO is available at http://www.nacc.usgcrp.gov.

National Assessment Coordination Office
Michael MacCracken,Lawrence Livermore National Laboratory, executive director
Melissa Taylor, University Corporation for Atmospheric Research, executive secretary of NAST (through March 2000)
Lynne Carter, University Corporation for Atmospheric Research, regional liaison
Robert Cherry, University Corporation for Atmospheric Research,administrative assistant (through March 2000)
Nakia Dawkins,AA Temps (from November, 1999)
LaShaunda Malone,University Corporation for Atmospheric Research, research associate and, from Sept.1999,sectoral liaison
Katherin Slimak,University Corporation for Atmospheric Research, research assistant (summers of 1998,1999)
Justin Wettstein,University Corporation for Atmospheric Research,sectoral liaison (through Aug.1999)

Assistance from Other Organizations
Joy Colucci,TerraComm,newsletter editor for Acclimations
Benjamin Felzer, National Center for Atmospheric Research,coordinator for climate scenarios
Susan Henson,National Science Foundation, NAST travel and SGCR administrative support
Forrest Hoffman,Oak Ridge National Laboratory,Web support
Lynn Mortensen,University Corporation for Atmospheric Research,education and outreach specialist
Mary Ann Seifert,Marine Biological Laboratory, administrator for the NAST co-chair

Technical and Expert Reviewers:

The NAST is particularly grateful for the over 300 individuals who provided very thorough reviews and provided many helpful comments.

GRAPHICS AND PHOTOGRAPHY SOURCES

Cover: NASA
Table of Contents:
Pages 2- 3: NASA
Pages 4- 5: NASA
Page 7: NASA
Page 9:1.Benjamin Felzer, UCAR;2.B. Felzer, UCAR;3.©Paul Grabhorn;4. ©P. Grabhorn;5.©P. Grabhorn;6.©P. Grabhorn;7. NASA;8.©P. Grabhorn; 9. NASA;10. NASA.
Page 10& 11:Basemap ©Clifford Grabhorn;Coastal Communities photo - Associated Press Laserphoto; Extreme Events – NOAA;all other photos - ©P. Grabhorn.
Page 12:Illustration - Melody Warford
Page 13:Temperature data from - from M.Mann,et.al.,(1999) Geophysical Research Letters;CO2 Data from –M.Etheridge,et.al., (1998) and C. D. Keeling (1999) through the Carbon Dioxide Information Analysis Center (CDIAC), Oak Ridge National Laboratory (ORNL);Carbon Emissions Data from - R.Andres,et.al.,(2000),R.Houghton, (1995,1996) and G. Marland,et. al.,(1999) through the CDIAC/ORNL..
Page 15:Based on illustration of National Climate Data Center (NCDC).
Page 17:all – B. Felzer, UCAR.
Page 18:Based on data from NCDC.
Page 19:El Niño graphics - NOAA Pacific Marine Environmental Lab home page http://www.pmel.noaa.gov/toga-tao/el-nino/nino-home.html#
Page 20:all – B. Felzer, UCAR.
Page 21:all – B. Felzer, UCAR.
Page 22:all – B. Felzer, UCAR.
Page 23:all – B. Felzer, UCAR.
Page 24:all ©P. Grabhorn
Page 25:photo - ©P. Grabhorn; Distribution of Plant Communities - based on graphic that first appeared in R.H.Whittaker, (1970). Communities and Ecosystems. Macmillan.New York.
Page 26:all ©P. Grabhorn
Page 27:Changes in Vegetation

Carbon - output of the Terrestrial Ecosystem Model (TEM) - run as part of the VEMAP II study – J. Melillo,et. al.,(1999) The Ecosystems Center, Marine Biological Laboratory,Woods Hole,MA.;all photos - ©P. Grabhorn.
Page 28 - 29:Ecosystem Models - output of the Mapped Atmosphere-Plant-Soil System (MAPSS) Model,R. Neilson,et.al.,(2000) US Dept of Agriculture(USDA), Forest Service - abstract appears AGU,Washington DC, May-June.
Page 30 - 31:21st Century Growth charts – P. Grabhorn,based on projections by Pacific Northwest National Laboratory (PNNL) and NPA Data Services,Inc.,in collaboration with the National Assessment Synthesis Team (NAST);City photo - ©P. Grabhorn
Page 32:US Population and Growth Trends - CIESIN, Columbia University, based on projections by NPA Data Services,Inc.,in collaboration with NAST.
Page 35:Hurricanes and their Impacts:updated by R.A.Pielke Jr., from P.J. Herbert,, J.D. Jarrell,and M. Mayfield,(1996).The Deadliest, Costliest,and Most Intense Hurricanes of this Century. NOAA Technical Memorandum NWS TPC-I.
**Page 36:California red tide photo - ; Florida and Maine plankton bloom satellite photos -
Page 37:Illustration by Bill Baker and Paul Grabhorn with content from NAST.
**Page 38 - 39:Annual Temp charts – B. Felzer, UCAR;Alaska Map - ;Islands Map - ;US Map - ©C.Grabhorn.
Page 40:Map - ©C.Grabhorn
Page 41:Photos – Mt.Washington, Barry Rock,University of New Hampshire;Temp.and Precip Maps – B. Felzer, UCAR;NY Storm Surge Map - Klaus Jacobs,Lamont-Doherty Earth Observatory of Columbia University.
**Page 42 - 43:Map - ©C.Grabhorn; Snow Photo - American Red Cross;

Chesapeake Satellite Image - NASA Goddard Space Flight Center, Scientific Visualization Studio, SeaWIFS image;Temp.and Heat Index Maps – B. Felzer, UCAR; Percent Salinity Change - J. Gibson and R. Najjar, Penn State University.
**Page 44 - 45:Map -©C.Grabhorn; Skier Photo - ;Stream Photo - ©P. Grabhorn;Dominant Forest Types Maps - Prasad,A.M.and L.R.Iverson. (1999-ongoing).A Climate Change Atlas for 80 Forest Tree Species of the Eastern United States [database]. http://www.fs.fed.us/ne/delaware/atl as/index.html,Northeastern Research Station,USDA Forest Service, Delaware,Ohio.
Page 46:Map - ©C.Grabhorn;Ghost Forest Photo - USGS.
Page 47:Coastal Loss Map - US Geological Survey (USGS) and Louisiana Department of Natural Resources;Temp.and Precip.Maps - B Felzer, UCAR.
Page 48:Map - ©C.Grabhorn;Flooded Community Photo - V. Burkett - USGS; Burned House photo - USGS.
Page 49:Crop Yields - Auburn University, Global Hydrology and Climate Center, University of Florida, Agricultural and Biological Engineering Department.
Pages 50 - 51:Map - ©C.Grabhorn; Forest Maps - USDA Forest Service, Southern Global Change Program; Timberland Acreage – North Carolina State University, Department of Forestry;Research Triangle Institute Center for Economics Research.
Pages 52 - 53:Map - ©C.Grabhorn; Temp.and Precip.Maps – B. Felzer, UCAR;Barge photo - ©P. Grabhorn.
Pages 54 - 55:Map - ©C.Grabhorn; Lake Ice Duration - John Magnuson, University of Wisconsin; Summer Climate Shift Maps - Don Wuebbles, University of Illinois and compiled by Byron Gleason NOAA,NCDC.
Pages 56 - 57:Map - ©C.Grabhorn; Midwest Soybean Yield - Dave

GRAPHICS AND PHOTOGRAPHY SOURCES

Easterling NCDC;Midwest Daily Precip.- compilled by Byron Gleason NOAA,NCDC.

Pages 58 - 59:Map - ©C.Grabhorn; Temp.and Precip.Maps – B. Felzer, UCAR;Tractor photo - ©P. Grabhorn.

Pages 60 - 61:Map - ©C.Grabhorn; Consumptive Water Use and Irrigated Water Use Graphics – Ojima,et.al., (1999). Potential climate change impacts on water resources in the Great Plains. JAWRA,35,1443-1454; Palmer Drought and July Heat Maps – B. Felzer, UCAR.

Pages 62 - 63:Map - ©C.Grabhorn; Spotted Knapweed,Leafy Spurge,and Yellow Starthistle Photos - http://www.blm.gov/education/weed /photos.html;Leafy Spurge Map - http://www.nhq.nrcs.usda.gov/land/i ndex/GoH.html;Soil Moisture Map – B. Felzer, UCAR;Soil Carbon & NPP Maps - Century results from VEMAP analysis,Natural Resource Ecology Lab,Colorado State University,

Pages 64 - 65:Map - ©C.Grabhorn; Population Chart - U.S.Census Bureau,(1998), California Trade and Commerce Agency, (1997);California Department of Finance,(1998),CLI-MAS,(1998),NPA Data Services,Inc., (1999);Temp.and Precip.Maps – B. Felzer, UCAR; Forest Fire photo - ©P. Grabhorn.

Pages 66 - 67:Map - ©C.Grabhorn; Water Use Chart – based on Solley, et. al.,(1998),Diaz and Anderson,(1995), CLIMAS,(1998),Templin,(1999); Ecosystem Maps - output of the MAPSS model,Neilson,et.al.,(2000) USDA Forest Service - abstract appears AGU,Washington DC,May-June.

Pages 68 - 69:Map - © C.Grabhorn; Reservoir photo - Seattle Public Utilities Department;Temp and Precip.Maps – B. Felzer, UCAR.

**Pages 70 - 71:Map - ©C.Grabhorn; Columbia Streamflow - Mote, et.al.,(1999) Impacts of climate vari-ability and change in the Pacific Northwest.University of Washington

– Summary;Climate Variability On Salmon - ;Regional Impacts Chart – P. Grabhorn adapted from Mote, et.al.,(1999) Impacts of climate var-ability and change in the Pacific Northwest.University of Washington.

Pages 72 - 73:Map - ©C.Grabhorn; Soil Moisture Maps – B. Felzer, UCAR; Forest photo - ©P. Grabhorn;Winter Snow Maps – B. Felzer, UCAR; Mountain photo - ©P. Grabhorn; Projected Northwest Daily Precip Change - NCDC.

**Pages 74 - 75:Map - ;Iceberg Photo - ©P. Grabhorn;Observed Temp and Precip - Data from Historical Climate Network,NCDC;Precip.and Temp. Maps – B. Felzer, UCAR.

**Pages 76 & 77:Map - ;Sea Ice Change Maps – P. Grabhorn,based on polar projections of Canadian model results by B. Felzer, UCAR;Boreal Forest Fire Chart - after Figure 1.2,pg. 2,in E.S.Kasischke and B.J. Stocks (eds.) (2000), Fire,Climate Change, and Carbon Cycling in the Boreal Forest,Ecological Studies Series,New York:Springer-Verlag,in press.Data from Alaska Fire Service,Canadian Fire Service; Fishing Boat photo - ©P. Grabhorn.

**Pages 78 - 79:Map - ;Soil Moisture and Winter Temp Maps – B. Felzer, UCAR;Caribou and Seal photos - ©P. Grabhorn;Vegetation Distribution Maps - Neilson,R.P.,I.C.Prentice, B.Smith,T.G.F. Kittel,and D.Viner (1998).Simulated changes in vegeta-tion distribution under global warm-ing,Annex C in R.T.Watson, M.C.Zinyowera,R.H.Moss and D.J. Dokken,(eds.),The Regional Impacts of Climate Change:An Assessment of Vulnerability. Special report of IPCC Working Group 2.Cambridge University Press,pg.439-456.

**Pages 80 - 81:Maps - ;El Niño Billboard photo - U.S.National Weather Service, Pacific Region Office.

**Pages 82 - 83:Beach photo - ©P. Grabhorn; Freshwater Lens - illustra-

tion by Melody Warford ;Hurricane Georges Map - ;Hakalau photo - ; Coral Reef photo - ©P. Grabhorn; Mangrove photo - ©P. Grabhorn;El Niño Charts - James O'Brien,Florida State University compiled by T Karl;

**Pages 84 - 85: Petroglyph photo - ©P. Grabhorn;Harvesting wild rice photo - .

**Pages 86 - 87:Indian ruins photo - ; Irrigable Indian lands - from "Atlas of the New West," W.W. Norton and Company, 1997;Salmon photo - , Landscape photo - ,Indian ruins photo - .

Pages 88 - 89:All photos - © P. Grabhorn.

Pages 90 - 91: Tractor photo - ©P. Grabhorn;Yield Charts – Changing Climate and Changing Agriculture: Report of the Agricultural Sector Assessment Team (2000).

Pages 92 - 93: Farm Photo - ©P. Grabhorn;Yield Charts,Regional Production Charts,and Economic Impacts Charts - Changing Climate and Changing Agriculture:Report of the Agricultural Sector Assessment Team (2000).

Pages 94 - 95:Flooded Farm photo - ©P. Grabhorn;Corn Yield Chart - Changing Climate and Changing Agriculture: Report of the Agricultural Sector Assessment Team (2000); Ranch photo - ©P. Grabhorn.

Pages 96 - 97: Water photo - ©P. Grabhorn;Water Withdrawal Chart – based on Solley, et.al.,(1998) USGS Circular 1200.Estimated Use of Water in the United States in 1995; Snoqualmie Pass Illustration – M. Warford;Columbia Basin Snow Extent – Mote,et.al.,(1999) Impacts of cli-mate variability and change in the Pacific Northwest,University of Washington;Winter Snow Cover Maps – Redrafted from data presented in McCabe, G.J.;Wolock, D.M.(1999) General-circulation-model simulations of future snowpack in the western United States. JAWRA,v 35,1473-1484.

Pages 98 - 99: Wetland photo - ©P. Grabhorn; Palmer Drought Maps – B. Felzer, UCAR, Observed Changes Chart - NCDC.

**Pages 100 - 101: Flooded House photo - ©P. Grabhorn; National Daily Precip. Chart – NCDC; Summer Stream Temp. Chart -; Prairie Pothole photo - ; Groundwater Recharge photo – Orange County Water District; CAP Canal photo - ©K. Jacobs; Subsidence photo - ©K. Jacobs.

Pages 102 - 103: Boys and Water photo - ©P. Grabhorn; Heat Index Maps – B. Felzer, UCAR; Heat Related Deaths - Chicago – NOAA; Present and Projected Heat Deaths Chart - Data from Kalkstein and Greene (1997). The three climate scenarios used in the study were GFDL, UKMO, and Max-Planck, 1985 runs using the IPCC transient scenario without aerosols.

Pages 104 - 105: NY City photo - ©P. Grabhorn; Maximum Daily Ozone Chart –EPA; Atlanta photos - ©P. Grabhorn.

Pages 106 - 107: Albuquerque photo - ©P. Grabhorn; Dengue Map - data from Mexican Ministry of Health and CDC; Potential Health Effect Chart - Patz et al., 2000

Pages 108 - 109: Coastal photo - ©P. Grabhorn; Coastal Vulnerability Map - USGS, Coastal Geology Program; Ocean Circulation Chart - Illustration by Melody Warford adapted from Broecker (1991) Nature 315:21-26; Sea Bird photo - ©P. Grabhorn.

Pages 110 - 111: Coral Reef photo - ©P. Grabhorn; Marsh Elevation Illustration - Illustration by Melody Warford based on original figure by USGS, National Wetlands Research Center; US Coastal Lands at Risk Chart – US EPA (1989). The Potential Effects of Global Climate Change on the United States. EPA # 230-05-89-050; Temp and CO2 Stresses Data Chart –Wiley Encyclopedia of Global Environmental Change (in press) and Kleypas et.al., (1999) Science 284:

118-120 with background photo - © P. Grabhorn; Bleached Coral photo – NOAA photo library; Calcium carbonate maps - Kleypas, J.A., R.W. Buddemeier, et al. (1999). Geochemical consequences of increased atmospheric carbon dioxide on coral reefs. Science, 284 (2 April 1999):118-120. The Reebase reefs were not shown on that map, but the juxtaposition of those reefs to the saturation state data was portrayed and discussed in Kleypas, J.A., J.W. McManus, et, al., (1999). "Environmental limits to coral reef development: Where do we draw the line?" American Zoologist 39(1):146-159.

Pages 112 - 113: Storm photo - Associated Press Laserphoto; Sea Level Rise Projections and Maps – B. Felzer, UCAR; Annual Shoreline Change Map - USGS, Coastal Geology Program.

Pages 114 - 115: Forest photo - ©P. Grabhorn; Current Distribution Map - Data are from USDA Forest Service http://www.srsfia.usfs.msstate.edu/rpa/rpa93.htm; Forest Land Percentages Chart - Data are from Forest Service Resource Bulletin PNW-RB-168, Forest Resource Report No.23, No.17, No.14, the Report of the Joint Committee on Forestry, 77th Congress 1st Session, Senate Document No.32. Data for 1850 and 1870 were based on information collected during the 1850 and 1870 decennial census, data for 1907 were also based on the decennial census modified by expert opinion, reported by R.S. Kellog in Forest Service Circular 166. Data for 1630 were included in Circular 166 as an estimate of the original forest area based on the current estimate of forest and historic land clearing information. These data are provided here for general reference purposes only to convey the relative extent of the forest estate in what is now the US at the time of European settlement (USDA Forest Service in review).

Pages 116 - 117: Foliage photo - ©P.

Grabhorn; Forest Fire Fighter photo - ©P. Grabhorn; Dominant Forest Types Maps - Prasad, A.M. and L.R. Iverson. (1999-ongoing). A Climate Change Atlas for 80 Forest Tree Species of the Eastern United States [database]. http://www.fs.fed.us/ne/delaware/atlas/index.html, Northeastern Research Station, USDA Forest Service, Delaware, Ohio.

Pages 118 - 119: Tree photo - ©P. Grabhorn; Change in Forestry Welfare Chart and Projected Average Timber Price Chart - Irland, L.C., D. Adams, R. Alig, C. J. Betz, C. Chen, M. Mutchins, B.A. McCarl, K. Skog, and B.W. Sohngen (2000). Assessing socioeconomic impacts of climate change on U.S. forests, wood product markets, and forest recreation, Bioscience, (in press); Tree photo - ©P. Grabhorn.

Pages 126- 127: Earth Image – NASA; Water photo - ©P. Grabhorn.

Pages 128 - 129: Maps - ©C. Grabhorn; Invasive Species photos - http://www.blm.gov/education/weed/photos.html; all other photos - ©P. Grabhorn.

Page 131: Hurricane Image - NASA.

Page 141: Earth with Hurricane Image - NOAA.

Scientific Writing and Editing:
Susan J. Hassol
Design, Layout and Production:
Grabhorn Studio